Derrida and Technology: Life, Politics, and Religion

Björn Sjöstrand

Derrida and Technology: Life, Politics, and Religion

Translated by Stephen Donovan

Springer

Björn Sjöstrand
Södertörn University
Stockholm, Sweden

This work was originally published in Stockholm under the title *Att tänka det tekniska: En studie i Derridas teknikfilosofi* by Södertörn Philosophical Studies (2015).

ISBN 978-3-030-83406-7 ISBN 978-3-030-83407-4 (eBook)
https://doi.org/10.1007/978-3-030-83407-4

This Springer imprint is published by the registered company Springer Nature Switzerland AG
The registered company address is: Gewerbestrasse 11, 6330 Cham, Switzerland

Preface

This book seeks to understand the significance of technology for Jacques Derrida's philosophical thinking. Strangely, this aspect of his writing has only very rarely been the object of philosophical analysis. What makes this omission stranger still is that deconstruction, as Derrida saw it, is closely connected to technology. As his entire body of writing makes clear, deconstructive thinking is deeply engaged with technology and its relation to human beings—and thus to life, ethics, politics, democracy, and religion. Derrida takes technology very seriously indeed, something that is far from always the case, particularly in philosophy, where philosophers typically prefer to spend their time analysing words, concepts, and ideas. He did so, moreover, at a time when technology, including in its concrete forms, has come to play a fundamental role in our experience and understanding of the world, the choices we make, and the decisions we take. The question of technology may have been largely avoided by philosophers, but we can nonetheless see how it has steadily grown in importance from the nineteenth century and, in particular, during the last hundred years. Understanding how technology emerges as a philosophical question and how Derrida addresses it is the primary purpose of this book.

The origin and inspiration for my choice of subject was a seminar series about issues in philosophy of technology that was convened by the Department of Philosophy at Uppsala University and the Philosophy research area at Södertörn University in 2006. Its thought-provoking lectures and discussions piqued my interest in this field, particularly as I had previously written a thesis on Derrida's views on ethics and politics.

Writing a philosophical monograph on so large and complex a subject as Derrida and technology would not have been possible without the help and support of colleagues and friends. I wish to record here my warmest thanks to each and every one of them. My first debt of gratitude is to Professor Sharon Rider at Uppsala University and Professor Hans Ruin at Södertörn University. Sharon has been an unfailingly steady support to me during the entire project, and her critical reflections and requests for greater clarity and stringency have greatly helped to focus my arguments. I also want to extend my heartfelt thanks to Hans for his faith in my project and for his tireless engagement, without which this book could not have been completed in its current form.

I have also incurred substantial debts of gratitude to all of my colleagues who read and commented on the book's various chapters. Special thanks to the participants in the Södertörn Higher Seminar in Philosophy for many stimulating discussions, particularly Marcia Sá Cavalcante Schuback, Sven-Olov Wallenstein, Fredrik Svenaeus, Fredrika Spindler, Nicholas Smith, Charlotta Weigelt, Mats Rosengren, Anders Bartonek, Carl Cederberg, Erik Bryngelsson, Gustav Strandberg, Jens Cavallin, Krystof Kasprzak, Lovisa Håkansson, and Ramona Rat. And likewise many thanks to the participants in the Uppsala Higher Seminar in Language and Cultural Philosophy, particularly Sören Stenlund, Erik Boström, and Johan Boberg.

My thinking about technology has been profoundly shaped by my professional background as an electrotechnical engineer with more than 30 years' experience of information and communications technology, by discussions with Derrida experts such as Bernard Stiegler, Jean-Luc Nancy, Robert Bernasconi, and by participation in international conferences, especially those arranged biannually by the Society for Philosophy and Technology, a worldwide organization for philosophers of technology with a range of backgrounds and interests. It has been a pleasure taking part in these conferences and, in particular, talking with Carl Mitcham and Andrew Feenberg about philosophy of technology in general and with Don Ihde and Peter-Paul Verbeek about postphenonomenology and technology.[1]

Lastly, I would like to acknowledge my friends who helped to rouse me from my "academic" slumber. I have benefitted from detailed and valuable commentary on the book's technical chapters from my former colleagues at Ericsson, Lars-Erik Lögdberg and Stefan Norrman. And a special thank-you to my good friend Mats Paulson, who, giving the inspiration that only an artist and a musician can, has been a source of intellectual stimulus and pleasure during our almost daily "philosophical walks".

Stockholm, Sweden Björn Sjöstrand
June 2021

[1] This translation of the original version has been updated to include secondary literature of importance in the past 5 years.

Contents

Chapter 1
Introduction

How should we understand technology and its relation to humanity? Where does technology end and humanity begin? How does technology relate to thought? In recent decades, philosophers have become increasingly interested in these and similar questions about *technology*. The origins of this growing interest unquestionably lie in the frenetic technologization of modern society. Explosive technological development and an accompanying globalization of economics and politics beyond the borders of the nation state now pose a challenge to humanity, its identity, and its self-conception. New technologies of production and reproduction, which range from the individual to the global level, are becoming increasingly intertwined with human life. Increasingly rapid technological developments, for example, in genetics, artificial intelligence, and cognitive research, compel us to ask the question "What is a human being?" with renewed urgency, while new media and communication technologies steadily encroach upon our lifeworld, particularly our understanding of time and space. This accelerating technological development also actualizes questions about the ethical, political, and religious consequences of technology.

For Jacques Derrida, these were fundamentally philosophical questions that required a radically new way of thinking. I wish to argue that his philosophy of technology, as outlined in this study, is that new way of thinking. Not only is it comprised of original reflections upon sophisticated artifacts and their role in our technological culture, it enables us, above all, to grasp the scope of what he calls "*our technological situation.*"[1] Indeed, it could even be said that, from 1967 onwards, he developed what he called *deconstruction* within the framework of this "situation" defined by unparalleled advances in technology. In a memorial lecture to his close friend Paul de Man, he underlines the close connection between deconstruction and technology: "There is no deconstruction which does not begin… by

[1] Derrida, *Points, 1974–1994,* pp. 244–5.

B. Sjöstrand, *Derrida and Technology: Life, Politics, and Religion*,
https://doi.org/10.1007/978-3-030-83407-4_1

calling into question the dissociation between thought and technology… however secret, subtle, sublime or denied it may be."[2] In this passage, deconstruction calls into question a Heideggerian philosophy of technology that relies precisely on the separation of philosophical thinking and technology itself. For Derrida, this thinking is always already contaminated by technology.

The aim of this study is to sketch the main features of Derrida's *alternative* philosophy of technology, a philosophy that, as I see it, avoids the problems associated with both the Heideggerian understanding of technology and the empirically-oriented "postphenomenology" that today dominates the field. I sketch the outlines of a coherent philosophy of technology, a *phenomenology of technology*, which radically extends the concept of technology to include the entire phenomenological field: technology is something that exists not in opposition to humanity but in a close relationship to humanity and hence to life, time, ethics, politics, democracy, and religion.

In this study, I seek to show that Derrida's entire oeuvre, from beginning to end, is permeated by reflections on *technology*. I establish the ways in which advances in modern technology served as a key source of inspiration for his repeated examination of genetic engineering, computer technology, teletechnology, archival technology, and media in all its forms. I show how Derrida, by engaging in a lifelong dialogue with the writings of Edmund Husserl and Martin Heidegger in particular, came to view technology not as a free-standing apparatus but as a *relation* between two "apparatuses" that resemble each other: the psychic apparatus and the non-psychic apparatus. In this relation, the latter constitutes a *supplement* that is necessary if the former is to be able to function at all. On the basis of this "definition," I then show that, for Derrida, technology is closely connected to humanity and its world: to thought and time, to life and death, and, above all, to ethics, politics, and religion. For example, he argues that technology will transform the very notion of the *political* as we know it. Modern technology's ability to compress time and space and spread ideas, sound, and images across the world at lightning speed will obliterate the boundaries between nation states, a fact that forces us to think the political in a new way, to think the political *beyond* politics. It is this *beyond*, which will liberate politics from its own identity and connect it with that which is entirely other, that Derrida calls "ethics."

For Derrida, *religion* is also intimately connected to technology. In the essay "Faith and Knowledge" (*Foi et savoir*), he shows that the same originary and elementary faith serves as the foundation for religion as well as science and technology.[3] He also shows that the relation between them is *autoimmune*, which paradoxically means that religion exploits technology *even while* rejecting it. This relationship of autoimmunity is perhaps most visible in religion's attempt to protect life, the sacred life, by relying on the very technological resources against which it is reacting with all its strength.

---

[2] Derrida, *Memoires for Paul de Man*, p. 108.

[3] Derrida, *Foi et savoir* ("Faith and Knowledge") §37.

Of particular interest is Derrida's view of the relationship between religion and *media*. He strongly asserts that religion today is indistinguishable from the media that spread the religious message around the world. This *medialization* of religion is, as he sees it, fundamentally a Christian phenomenon. There is, he argues, a feature that is absolutely unique to Christianity: its universal claim and its desire to use the media to spread the Christian gospel around the world. Christianity, to a far greater degree than other religions, is therefore a *religion of the media*.

Surprisingly, these and other of his reflections on technology have only rarely been analysed or discussed by his many commentators. One exception is the French philosopher Bernard Stiegler, who was the first to propose that deconstruction should be regarded as a philosophy of technology. Yet Stiegler's unorthodox interpretation of Derrida's thinking eventually resulted in what might be called a "deconstructive materialism," a form of technological determinism that was alien to Derrida.

Apart from its appearance in Stiegler's writings, Derrida's name has been largely absent from debates among philosophers of technology and anthologies on the philosophy of technology. Nor has there been any overarching consideration of his ideas about technology beyond a handful of texts, published at various times, that address specific aspects of his thinking on these issues.

Peter Kemp's *Döden och maskinen* (Death and the Machine), which was published back in 1981, is a readable but rather dated introduction to Derrida. It was originally intended as "a critique of Derrida for having avoided ethics."[4] In its fourth and final chapter, whose subject is precisely death and the machine, Kemp criticizes his thinking on the machine for its lack of an ethical dimension. Yet this criticism must be regarded as unfounded, particularly in light of Derrida's subsequent writings. As will become clear in Chap. 6, "Ethics and Technology," the reverse is true: the ethical is continually present in deconstruction. Kemp's critique is similarly baseless when he assumes that Derrida's philosophy presupposes the existence of a "Technology as such" whose only problem is how we use it. On the contrary, Derrida argues, as does Kemp, that no such monolithic Technology exists. Instead, there are only different technologies, which must be analysed and evaluated individually with regard to their potential for promoting democratic development.

After publishing his book in 1981, Kemp discovered that deconstruction was in fact concerned with the ethical. In an epilogue appended to the Swedish translation, which appeared in 1990, he foregrounded—he was, perhaps, the first to do so—the fact that deconstruction makes ethical claims: "Derrida's latest work of philosophy has thus answered the question we have been asking in this book about the *danger of the machine*. … He now states that the deconstruction of our concepts, which makes us open to the unforeseeable, is itself an ethical task."[5]

The first book to thoroughly address the political implications of Derrida's philosophical project was Richard Beardsworth's *Derrida and the Political* (1996). In the

[4] Kemp, *Döden och maskinen*, p. 9.

[5] Ibid., p. 192.

closing pages of his conclusion, Beardsworth sketches the outlines of "two possible futures" for deconstruction. He describes the first as a 'left-wing' 'Derrideanism' that emphasizes the deconstruction of an "originary technicity." This alternative forms the philosophical background of Bernard Stiegler's philosophy of technology. The other possible future for deconstruction, which Beardsworth calls 'right-wing' 'Derrideanism', claims that the messianic *promise* is the very core of philosophical thinking.[6] Choosing this alternative, Beardsworth argues, requires mobilizing a religious discourse and passively awaiting an unforeseeable future, an uncertain coming. However, as I will show in Chap. 7, "Politics and Technology," there is no convincing basis for Beardsworth's "loose speculations" about such a dualism between a technological deconstruction and a religious deconstruction.

A subsequent text on the same topic is Philip Armstrong's short essay "Rethinking the Political" from 2007. In it, Armstrong discusses a passage in *Échographies de la télévision* in which Derrida argues that the concepts of politics, democracy, and citizenship are being radically changed by the limitless teletechnologies of our own era, and that the political today is essentially bound up with these technologies.[7] According to Armstrong, this fact necessarily requires us not only to think the political in terms of more than the borders of the nation state, but to rethink "the political beyond politics" or the "democratic beyond democracy," what Derrida calls "the democracy to-come." Adding very little of his own, Armstrong follows Derrida in rejecting the idea of the democracy to-come as a kind of "technological community" that tends to exclude large swathes of the world's population. Instead, like Derrida, he insists upon the French word *partage* or "sharing," which in itself designates both what is possible within a group and the fact that many people or groups in different parts of the world "have access to the same programmes."

David Wills's original studies *Prosthesis* (1995) and *Dorsality* (2008) seek to understand not only human nature prior to any technology but, above all, what we mean by technology. In both works, Wills refers at several points to Derrida's prosthetic thinking as it appears in certain passages in *Glas*, *Politiques de l'amitié*, and *La carte postale*, which Wills further develops in interesting directions.[8] When he states that we must begin "to think the subject of love and friendship, erotics and politics, in its biotechnological becoming," his prosthetic thinking is entirely in line with Derrida's.[9]

The absence of any systematic examination of Derrida's thinking on the relation of ethics and politics to technology also holds for his thinking on *religion*. To date, there has been no comprehensive and thorough study that does justice to his thinking on the relation of religion to technology. Michael Naas acute and stylish *Miracle and Machine* (2012) is a kind of reader's guide to Derrida's most important essay on

---

[6] Beardsworth, *Derrida and the Political*, p. 156.

[7] Derrida, *Échographies de la télévision*, pp. 65–6. This passage is considered in depth in Chap. 7, "Politics and Technology."

[8] "Prosthesis" (from the Greek *pro,* "in addition to", and *thesis,* "arrangement") denotes an artificial device for the replacement or enhancement of human bodily functions.

[9] Wills, *Dorsality,* p. 159.

faith and knowledge, *Foi et savoir. Miracle and Machine*—whose title carries echoes of the phrase *deus ex machina* even as it offers a kind of translation of *Foi et savoir*—is Naas's way of expressing the idea that "the unexpected and the unforeseeable—even the miraculous … can come about only within or by means of a certain machine."[10] More specifically, Naas formulates three theses on "the relationship between religion, science, and the media … between religion and contemporary tele-technoscience and the processes of globalization that are facilitated by it."[11] Although Derrida himself does not expressly state these theses, they are implicit in his text, and his entire argument is built around them. These three theses are considered more closely in Chap. 8, "Religion and Technology."

Arthur Bradley's *Originary Technicity* (2011) is a readable engagement with philosophical anthropology and Continental philosophy of technology, from Marx to Derrida. Specifically, the book offers a critical survey of the literature on the concept of *originary technicity*, here taken to refer not only to the way in which the human subject creates and uses technology but how it is always already constituted by that technology. Or, as Bradley explains in his introductory chapter: "To state it very simply, I wish to argue that the theory of originary technicity from Marx to Derrida is—despite its claims to the contrary—*not technical all the way down*."[12] When Bradley critically examines this theory, he finds that in every case there remains a genie in the bottle, an irreducible vestige of an independent, self-constituting subject that refuses to be subordinated to an originary technicity. Bradley's radical anti-subjectivism is unquestionably the most problematic aspect of his book. After all, his own book has been created by a person, for people. We can never understand what a creative being would look like who was completely determined by its own technology and stripped of all subjectivity. This does not mean that we have to fall back upon pure idealism. When Bradley, in his chapter on Derrida, criticizes what he regards as an "idealistic remainder within Derrida's deconstruction," he is jumping to conclusions because of the problematic starting point for his critique, namely that we are the fully determined creators of the technology that determines us.[13]

Bradley levels the same criticism at the other thinkers whom he discusses, primarily Marx and Engels, Freud, Lacan, Heidegger, and Stiegler. Perhaps the most absorbing chapter in the book is a comparison of Stiegler's and Derrida's respective views on the relationship between technology and time, in which Bradley assigns Stiegler a more materialist position as someone who sees time not in phenomenological terms but as ultimately created by and through technology. The final study that warrants mention is Ben Roberts's essay "Stiegler Reading Derrida," which is a close reading of Stiegler's discussion of Derrida's *La technique et le temps* and Stiegler's interviews with him in *Échographies de la télévision*. Roberts shows

---

[10] Naas, *Miracle and Machine*, p. 7.

[11] Ibid., pp. 2–3.

[12] Bradley, *Originary Technicity*, p. 15.

[13] Ibid., p. 119.

convincingly that the substantial differences between their respective positions partly stem from Stiegler's problematic reading of Derrida. I will examine the origins of this problematic in greater detail in Chap. 4, "Technology and Time As Event," and in Chap. 7, "Politics and Technology."

These are basically the texts that have in any way taken up the question of Derrida's ideas about technology.[14] Some are dated, others problematic, and a few have made important contributions to Derrida scholarship. How is it that there has been no systematic study of this central theme in his writing? One reason is doubtless that he himself never attempted a systematic exposition of this topic, as he did with concepts such as justice, religion, or friendship. But if we comb his entire output for this topic, we nonetheless see a pattern of analyses and positions that collectively form a deconstructive approach to technology.[15] This partly implicit philosophy of technology is what I will be trying to reconstruct in the following chapters.

In my view, reconstructing this philosophy of technology is a central element in deconstructing the prevailing image of Derrida's thinking on language and text etc.,

---

[14] [Since the original version of this book was published in 2015, a couple of texts have appeared, two of which are of particular interest here:

(a) In "Jacques Derrida et le problème de la technique"(2016), Tacettin Ertuğrul discusses the relationship between the question of technique and the question of writing in Derrida's work. After a close reading of Plato's determination of writing as *pharmacon* and Rousseau's determination of culture as *supplement,* he describes technique as "pharmaco-supplementary." But he goes on to introduce the term *archi-technicity*, which he claims is closely associated with arche-writing. What is meant by *archi-technicity,* however, seems somewhat unclear.

(b) In an interesting article, "Derrida's Quasi-Technique" (2016), Susanna Lindberg describes Derrida's deconstruction of different forms of thinking technique as essence of technique. She shows that Derrida's alternative technique is neither empirical nor transcendental but can be described as a *quasi-technique.* This "quasi-concept" can serve as a "salutary 'deconstruction' of mainstream philosophy of technology (that 'knows' the 'essence of technology') because it shows how to think technique in the absence of essence and as the absence of essence."

(c) In addition to these two texts, a conference on the theme "Derrida et la téchnologie" was arranged by the Columbia Global Centre in Paris from 2 to 4 May 2019. Participants included Arthur Bradley ("In the Machine of Sovereignty: Sovereignty, Governmentality, Automaticity") and Hans Ruin ("Techno-logical Faith – On Security, Threat and Salvation in Derrida's Thinking of Technology"). However, as far as I know, no proceedings from the conference have been published. *B.S.*]

[15] My interest in this topic stems in part from my personal background as an engineer and in part from my early contact with Derrida's ethico-political writings. While working on my master's thesis, which examines Levinas's and Derrida's views on the relation between ethics and politics, I found that he also had interesting things to say about technology. In particular, his "Nietzsche and the Machine" made me realize that his treatment of technology was a subject in its own right. The more I immersed myself in his ideas, the more I discovered that not only does he write about what technology is—its relation to human beings, time, life, and so forth—he also repeatedly returns to the practical aspect: technology's relation to the ethical, the political, and the religious. What I found in his writing was not a single work that systematically addressed technology in particular but, rather, an original and innovate philosophy of technology that permeated his entire oeuvre, from beginning to end, a philosophy of technology that turned out to have been largely overlooked. This is the background to my attempt to foreground and analyse Derrida's ideas about technology.

in which the question of technology is suppressed and marginalized. My goal is precisely to deconstruct this image and call into question this hierarchy by foregrounding the other, i.e. Derrida's ideas about technology: to reconstruct his thinking on technology in order to thereby destabilize the current image of this thinking and allow new perspectives and insights to emerge. Without such a deconstruction, there is a danger that his thinking will be reduced to a narrow analysis of linguistic and other concepts.

However, this does not mean that deconstruction should be thought of as a "method." Like Derrida, I argue that deconstruction is not a method. It is not a mechanical process. Philosophers with a method have already decided on a routine, much like "civil servants" who have to follow sets of rules, unable to allow their thinking to extend to the unexpected, the completely other. For this reason, I argue, reconstruction of Derrida's technology of philosophy is likewise not a method.

However, it does presuppose a certain degree of technological understanding and, above all, a deeper understanding of *technology as a philosophical question*: how this question gradually emerged in history, and how it has been addressed in differing ways by the greatest philosophers. Reflections on technology, the philosophy of technology, is now a well-established philosophical discipline with its own departments, conferences, and journals.[16] While it admittedly remains a relatively marginal branch of philosophy, the scholarship on technology has nonetheless grown prodigiously since the middle of the last century and today represents an enormous and not easily surveyed body of writing. In an effort to manage this sprawling literature, scholars have sought to classify it in a number of ways. One such classification has been proposed by Carl Mitcham, who argues in *Thinking through Technology* (1994) that it is possible to distinguish two broad traditions of philosophy of technology: an engineering tradition and a humanist tradition. *The engineering tradition* is made up of various attempts by engineers and technicians themselves to devise a philosophy of technology by investigating technology from within, as it were, and by regarding a technological way of being in the world as paradigmatic for other ways of thinking and acting. *The humanist tradition* represents various efforts by humanists, particularly phenomenologists, to understand modern technology from a phenomenological-hermeneutic perspective. Philosophy of technology in the humanist tradition is a way for philosophers to reflect deeply on technology as a topic—one topic among many others. Whereas the engineering tradition tends to be more positive about technology, the humanist is generally more critical and hermeneutic.

While this categorization has something useful to say about different philosophical approaches, it comes with a few specific problems. One is that it tends to become overly schematic or simplified: philosophical thinking does not lend itself to pigeonholing. Throughout its history, leading thinkers have ranged freely across the entire field of philosophy. Another problem is that division into categories tends to be

[16] As will become clear in what follows, Derrida's philosophy of technology represents an original and innovative contribution to the larger debate within philosophy of technology.

over-ambitious and insufficiently complete: large numbers of philosophers have to be incorporated into the schema in order to justify its existence while many others receive no mention. A third problem is that categorization easily overlooks some philosophical approaches to technology. For example, Mitcham's schema could be supplemented with a *social sciences approach* that examines the relationship between technology and society from a social, political, economic, or cultural perspective. This approach would differ from the two others in treating technology as an integrated part of society more broadly.

An alternative schema could be to divide philosophy of technology into separate historical phases. One example would be how Peter-Paul Verbeek divides philosophy of technology into three historical phases in his pioneering book *What Things Do* (2005). The first, which is represented by its founders, Martin Heidegger and Jacques Ellul, stretched from approximately 1925 to 1955. It was followed by period of decline lasting about 20 years. The second phase, which began in the United States in the early 1970s, rescued philosophy of technology from its supposed errors and established "technology" as paradigmatic of contemporary culture. This phase has now concluded and is in the process of being replaced by a third generation, to which Verbeek himself belongs.

Yet this division is also associated with particular difficulties. It overlooks, for example, the sustained philosophical engagement with technology that took place within neo-Hegelianism, neo-Kantianism, positivism, and phenomenology between roughly 1850 and 1920. Another difficulty is that the historical phases overlap, blurring the boundaries between them. For instance, it is hard to see why Verbeek makes such a sharp distinction between Don Ihde's "postphenomenological" discourse and his own. However, perhaps the greatest difficulty—which affects all such attempts at classification—is that they are not completely neutral. They invariably tend either to describe philosophy of technology in terms of the writer's own thinking or to treat such thinking as the natural *telos* of history.

In this study, I instead try to describe how technology made its gradual historical emergence as a philosophical issue. I begin by summarizing the eighteenth-century reaction to the Industrial Revolution's negative impact on people's lives, and then show how this pessimistic view gave way to optimism about technology among Enlightenment thinkers who dreamed of a society based on technological-scientific principles. Yet it was not until the mid-nineteenth century that philosophers interested in technology began to publish works that focused specifically on the issue of technology. These philosophers were Left Hegelians such as Ernst Kapp and Karl Marx, but eventually neo-Kantians and other technologically-oriented philosophers also began asking questions about technology. The "engineering" approach adopted by these latter took its point of departure in technology itself, and they typically accepted technology or took it for granted. They also took an interest in technology's many applications, as a result of which their work came to span an enormous field.

With the outbreak of the First World War, technology emerged in force as an urgent philosophical issue. This urgency was reinforced by the extremely rapid expansion and general dissemination of technology during this period. By the end of the 1920s, new technologies had come to affect every aspect of human existence, particularly modern warfare, which ceased to be the preserve of armies fighting battles far removed from people's everyday lives. War had become a battle for life and death for ordinary people and, indeed, all society.

In view of the new military technology, colossal material destruction, and unimaginable loss of life on the battlefield during the First World War, it was hardly surprising that the leading philosophers of technology at the time chose to paint a pessimistic portrait of their times, one in which humanity has become not only technology's master but also its slave. Technology is no longer an aid for improving the conditions of people's lives; quite the opposite, humanity has become the means by which that technology expresses itself.

The issue of technology became even more acute in the wake of the dropping of atomic bombs on Hiroshima and Nagasaki at the end of the Second World War, followed by the realization that nuclear weapons threatened to destroy humanity entirely. It was in this bleak atmosphere that Heidegger published his critical and oft-quoted essay *The Question Concerning Technology* (1954), which describes how humanity is on the way to losing control of its own creations. Humanity no longer controls technology; rather, technology itself increasingly appears as an autonomous framework within which everything is interpreted. Human beings as "enframed" are manipulators and exploiters of others, or, conversely, defenseless victims of manipulation and exploitation. All of "life" has been *totally mobilized.*

Heidegger's essay soon came to be regarded as emblematic by other phenomenological, hermeneutic, and existentially-oriented philosophers, who described technology in various ways as an all-encompassing, autonomous, and alienating force beyond individual things—a kind of framework or condition of possibility for knowledge and meaning. Collectively, they painted a grim picture of technology's role in society, and they worried that technology would ultimately alienate people from themselves and from reality. They feared that an increasingly technologistic way of thinking would limit or constrain the lives of modern people, their imagination, and creative capacity and prevent them from seeing the world as valuable in itself. Instead, people would be directed to view the world as "raw material," a resource for technological manipulation. In this technology-saturated world, humans are no longer unique individuals but anonymous workers destined to maintain a streamlined production apparatus for mass consumption. Although this grim diagnosis is a perhaps understandable reaction to the radical transformation of society by rapid technological development, the fact of its inevitability has increasingly been questioned in light of the emergence and implementation of specific technologies such as the new information and communication technologies. For critics today, this philosophy of technology seems excessively abstract and indiscriminate: abstract because it does not consider concrete technological applications, and indiscriminate because it invokes "Technology" with a capital T, leaving barely any room for philosophical reflection on different kinds of concrete technologies.

From one perspective, this form of philosophy of technology can be described as transcendental, since, following Kant, it starts from an analysis of the condition of possibility.[17] It tries to understand technology on the basis of this condition of possibility, of what must be assumed for it to be possible. This approach has provided us with many valuable insights and has shaped to a considerable degree our understanding of technology and its role in our culture. But the image of technology becomes distorted if technology is understood *exclusively* as its condition of possibility, because we would then be talking about technology's condition of possibility in the same way as if we were talking about concrete technologies as such. When we talk about technology exclusively in terms of its premises, it becomes impossible to investigate the role played by different technologies in different settings.

Empirical studies of technology and science conducted by scholars in recent decades fundamentally challenge the image of technology as an all-encompassing and alienating force. By studying concrete applications of specific technologies, they have shown that these technologies are experienced differently in different contexts. In other words, technology's supposed determinism seems to be weaker than previously assumed. As a result of these new findings, several leading philosophers of technology have developed traditional phenomenology in a direction that, they argue, makes possible a philosophy of concrete technologies. This new direction, which has had a considerable, perhaps even dominant, influence in recent philosophy of technology, has been termed "postphenomenology" by its founder, Don Ihde. Idhe and his follower Peter-Paul Verbeek argue that this concept offers a more balanced image of modern philosophy as it is expressed in language theory and postmodernism. Postphenomenologists claim that a fully developed philosophy of technology must take into consideration "concrete" empirical reality but without abandoning the philosophical questions posted by the older philosophy of technology. They seek to refocus philosophical attention away from technology's conditions of possibility onto technology itself—the technical apparatuses and objects that surround us in our daily lives—as a way of better understanding their role in our society and culture. Despite giving concrete technologies and artefacts the attention they undoubtedly warrant, postphenomenologists are rarely successful in their reflections on the phenomenological, hermeneutic, or existential questions that are so central for the first kind of philosophy of technology.[18]

As has already been noted, Derrida's philosophy of technology avoids some of the problems associated with the phenomenologically-oriented philosophy of technology and the postphenomenological school. In order to position his alternative line of thinking more clearly, my opening chapter, "The Philosophy of Technology Tradition," outlines a more detailed account of the historical development than the one I have just given.

In Chap. 3, "Derrida and the Phenomenology of Technology," I examine in greater detail Derrida's lifelong engagement with Edmund Husserl's and Martin Heidegger's writings on technology. I describe how a dialogue with these texts

[17] Verbeek, *What Things Do*, pp. 7–8.

[18] For a critical analysis of postphenomenology, see Robert Scharff, "Varför heideggerianer kan älska sina laptops utan dåligt samvete" ["Why Heideggerians Can Love their Laptops without Guilt"], pp.73–97.

came to form the basis for his own reflections on technology. From the very start, Derrida was profoundly engaged with Husserl's phenomenology, and at an early stage he aligned himself with the latter's conviction that ideal objects have a *technological* origin. He also develops Husserl's idea that *writing* is a technological tool that makes possible the opening of a connection in time and space, a telecommunication by which "reconnective" questions can be posed to the scientific traditions.

It is this deep involvement with Husserl's texts that gradually leads to Derrida's understanding of technology. I show how, on the basis of these texts, he develops a new type of logic, *the logic of the supplement*, upon which he bases his thesis that technology is a relation between the psychic and the non-psychic apparatus. This "definition" of technology is obviously incompatible with Heidegger's view of technology, above all as articulated in his late work *The Question Concerning Technology*. Derrida recognizes Heidegger as one of the first thinkers to make a thoroughly philosophical examination of the big question of what technology actually *is*. At the same time, Derrida identifies serious problems that, as he sees it, are especially apparent in Heidegger's later writings of the 1950s and after: their anti-technological tendency, their reduction of technology to its essence, and their view that humans are ultimately spiritual beings, elevated above and cleansed of every kind of technology.

The fact that Derrida regards technology as the psychic apparatus's relation to the technological apparatus means that technology is in some way related to the phenomenon of time. In Chap. 4, "Technology and Time As Event," I therefore analyse the texts in which he addresses the issue of time. Starting with an analysis of normal chronological time, what Heidegger called "vulgar" time, I proceed to explore Derrida's relation to Heidegger's concept of time, particularly his elaboration of the late Heideggerian notion of time as *event*. For Derrida, time consists of a series of events that suddenly intervene in and interrupt the flow of normal, homogeneous time. Today more than ever, he says, these events are produced by a gigantic teletechnological apparatus that comprises not only media in the form of the press, radio, TV, and the internet but also systems of mobile communication, digital archiving systems, and their accompanying infrastructure at a global level. In other words, time is produced *artificially*; it is an artifact. Time is created by a modern teletechnology that dramatically reshapes our experience of time, a time that is not only being transformed at an accelerating rate but that is also in the process of losing its character as a homogeneous temporal flow. Time, as Hamlet observes, *is out of joint*.

In Chap. 5, I take up the question of humanity's technicity, that is to say, whether and, if so, to what extent humans should be regarded as technological beings. The current debate over *human enhancement* is shining a spotlight on the old question about the essence of humanity: *what is a human being?* In contrast to the prevailing view that human life and technology are diametrically opposed, Derrida holds that human life and life in general are always already *contaminated* by technology. In this chapter, I try to explain the implications of this contamination and Derrida's statement that life is an extremely general "programme"—*différance*—which encompasses a whole chain of evolutionary stages running from the most elementary mechanisms of genetic inscription all the way to alphabetic writing. The fact that humans can in some respects be described in "technological" terms does not mean, however, that he defines humans as solely and essentially technological-biological beings, any more than they are purely transcendental subjects.

The always-already inscription of technology in life also has consequences for our understanding of subjectivity. Derrida's subject can no longer be described as a centre of Being but must instead be understood as an "effect" within a system of writing that is open to the technological world around it. The human subject must be understood as a kind of "writing machine" that is continually inscribing and erasing *traces* of the world around it and that "constitutes" the subject through the movement in *différance*.

In the two chapters that follow, I examine the ways in which technology relates to the ethical and the political. In Chap. 6, "Ethics and Technology," I seek to show that the ethical dimension of Derrida's thought is closely bound up with technology. I begin the chapter with an analysis of his understanding of the concept of ethics.

As I see it, Derrida's philosophical orientation, which is highly influenced by Emmanuel Levinas's ideas about ethics, is ethical through and through. His writings are characterized by an uninterrupted engagement with the wholly other—*le tout autre*—an engagement that involves both an obligation to and an assumption of responsibility for *alterity*. He accordingly defines ethics as *openness to the wholly other*. Ethics demands openness to the wholly other, which can come at any time from an unknown future that can never be foreseen. For Derrida, this alterity, which comes suddenly, if and when it does come, is an *event*. As we saw earlier, the event, perhaps more than ever, is *technologically mediated*. To illustrate the connection between ethics and technology, I devote the bulk of the chapter to three events of world-historical importance: the electronic revolution, total nuclear war (as a hypothesis), and global terrorism.

In Chap. 7, I seek to show that politics is also closely bound up with technology, opening with a discussion of the close relation between politics and ethics, followed by a section on Derrida's view of the political. In particular, I highlight his view that modern technology will radically transform the political as it is currently understood. A technology that knows no limits will soon erase the borders of the nation state, forcing us to see the political *beyond* politics. It is this *beyond* that liberates the latter from its own identity and connects it to the wholly other, what Derrida calls "ethics." The concepts of technology, ethics, and politics thereby become inextricably intertwined with a global techno-ethical-political complex in which ever-faster technological advances are themselves the motor of a development towards what he sees as a new world order.

The eight and final chapter, "Religion and Technology," is essential, in my view, for understanding Derrida's philosophy of technology as a whole. Religion reveals itself as an aspect of the question of technology that provides the key to a more comprehensive understanding of Derrida's thinking on this issue. Following an introductory summary of the key ideas of the previous chapters, I offer a brief account of his views on religion. Derrida often refers to his deep interest in questions of religion even though he considers himself an "atheist" in the sense of not believing in the existence of an other-worldly supreme being, a metaphysical-philosophical super-ego. By contrast, several of his writings allude to a "religion without religion," an intra-worldly "religion" that does not need religious revelations. The question then arises: if we can speak of a religious "turn" in Derrida's later thinking, must we then regard deconstruction as a kind of religion?

Derrida's most important analysis work on religion is unquestionably *Foi et Savoir*, "Faith and Knowledge," which contains the reflections that he contributed to a conference on religion held on Capri in 1994. Interestingly, it addresses not only the themes that we would expect to find in a book on religion (faith, prayer, sacrifice, salvation, etc.) but also, and perhaps more surprisingly, issues such as teletechnology, telecommunications, globalization, and the media. The remainder of the chapter therefore focuses mainly on this key text about faith and knowledge, religion and science, and religion and technology. I begin by summarizing the circumstances of the Capri conference: the location and the delegates, their languages, nationalities, and religious views, all of which are essential for understanding Derrida's reflections in "Faith and Knowledge." I then examine his effort to "define" the nature of religion and the role played by religion in the contemporary world. In connection with this, I address the difficulty of speaking about religion today when the language we use is, perhaps from the beginning, already a language of religion, in other words, a Latinate language that defines so many of our religious concepts, including the concept of religion itself. Thereafter I discuss the main message of "Faith and Knowledge," which, following Naas, I propose can be summarized as three theses. The first identifies the two sources of religion. The second claims that an originary and elementary (undogmatic) faith forms the basis for religion as well as science and technology. The third thesis, which I argue is Derrida's main thesis, holds that the relation between religion and technology is not only essential but *autoimmune*.

In the following section, I specifically address Derrida's handling of the relationship between religion and the media. As noted earlier, Derrida firmly maintains that religion is indissociable from the media that are spreading the religious message around the world via books, the internet, radio, and TV, and that the medialization of the world is fundamentally a Christian phenomenon. Although medialization occurs in all religions, he argues, one aspect is absolutely unique to Christianity: the desire to spread the "good news" of the gospel. At this point he rather daringly advances the thesis that we are today experiencing the end of a Christianity that uses the media better and more effectively than all other religions, and the beginning of *a religion of the media*.

I conclude by returning to the question of a possible religious "turn" in Derrida's later thinking, undoubtedly one of the most fiercely debated topics within contemporary Derrida research. I approach this issue by analysing the debate between John Caputo and Martin Hägglund in terms of their respective views of deconstruction and its relation to religion and technology. Although their positions seem in many regards to be reverse mirror images of each other, both men apparently agree with Derrida on at least one point: there is a close connection between deconstruction and technology.

Deconstruction always begins by questioning the separation of thought and technology. It likewise questions any philosophy of technology that builds on the separation of philosophical thought and technology itself. It thereby also calls into question Heidegger's famous adage: "the essence of technology is not something technical." In this study, I show that the unique hallmark of Derrida's philosophy of technology—*phenomenology of technology*—is precisely its broadening of the concept of technology to include the entire phenomenological field: for him, it is impossible to detach thought from technology. Thinking is always already contaminated by the technological. We must take this contamination as a given. Without it, we cannot assume responsibility for the technologization of the world. Without it, we cannot open our minds to the future, its perils, and its possibilities.

References

Beardsworth, R. (1996). *Derrida and the political.* Routledge.
Bradley, A. (2011). *Original technicity: The theory of technology from Marx to Derrida.* Palgrave Macmillan.
Derrida, J. (1988). *Memoires pour Paul de Man.* Galilée.
Derrida, J. (1995). *Points...interviews, 1974–1994.* Edited by Elisabeth Weber, (trans: Kamuf, P., et al .). Stanford University Press.
Derrida, J. (2000). *Foi et savoir.* Seuil.
Derrida, J., & Stiegler, B. (1996). *Échographies de la télévision.* Galilée.
Derrida, J., & Stiegler, B. (2002). *Echographies of television: Filmed interviews.* (trans. J. Bajorek). Polity Press.
Ertuğrul, T. (2016). *Jacques Derrida et le problème de la technique.* Université de Strasbourg.
Kemp, P. (1990). *Döden and maskinen: En introduktion till Jacques Derrida.* [Death and the Machine. An introduction to Jacques Derrida]. (trans. Eriksson, U.I.). Symposion.
Lindberg, S. (2016). Derrida's quasi-technique. *Research in Phenomenology, 46*, 369–389.
Naas, M. (2012). *Miracle and machine: Jacques Derrida and the two sources of religion, science and the media.* Fordham University Press.
Scharff, R. (2011). Varför heidiggerianer kan älska sina laptops utan dåligt samvete. [Why Heideggerians Can Love their Laptops without Guilt]. In L. Dahlberg & H. Ruin (Eds.), *Fenomenologi, teknik and medialitet* (pp. 73–97). Södertörn University.
Verbeek, P.-P. (2005). *What things do.* (trans. Crease, R.P.). Pennsylvania State University Press.
Wills, D. (2008). *Dorsality: Thinking back through technology and politics.* University of Minnesota.

Chapter 2
The Philosophy of Technology Tradition

2.1 The Historical Background

The Greek word τέχνη, which is usually translated as "art," "craftsmanship," or "skill," is generally used to designate the human capacity for producing, making, or creating something. In classical philosophy, however, *techne* was regarded not only as a human activity in the broadest sense but, above all, as a kind of knowledge. In Plato, who was the first to examine the concept thoroughly, *techne* and *episteme*, art and knowledge, are closely connected. Even for Aristotle *techne* is a kind of knowledge that characterizes a certain form of human activity. In his view, the individual craftsman aims to achieve as faithful a representation of nature as possible despite being fully aware of the impossibility of creating exact copies of his own work. *Techne* in this classical sense—and this cannot be emphasized enough when comparing ancient and modern technologies—is thus fundamentally directed towards unique and individual artifacts, not the efficient production of multiple objects for the purposes of making money. For the ancient craftsman, mass-production would have been unthinkable, and not just for technological reasons.

In the early seventeenth century, however, the classical view of technology was dramatically transformed. It was at this time that the classical worldview collapsed, and modern science first saw the light of day. The threshold figure in this scientific revolution was Galileo Galilei. Galileo gave impetus to the new science when he used his new telescope—which was a considerable improvement on Hans Lippersheys's invention—to look out at the cosmos. After his death in 1642, doctors, astronomers, physicists, and philosophers all began to adapt their work to Galileo's principles. Without enquiring into classical dogmas and authorities, they began investigating reality directly by using scientific experiments to try to uncover its secrets—weighing, measuring, counting, and devising a host of ingenious instruments. They used barometers and air-pumps to analyse the properties of air, and the microscope revealed to them previously unknown worlds within drops of water and

B. Sjöstrand, *Derrida and Technology: Life, Politics, and Religion*,
https://doi.org/10.1007/978-3-030-83407-4_2

the human body. Descartes sketched the outlines of a universe governed by inexorable mechanical laws. The crowning achievement of this development came near the end of the seventeenth century with the publication of Isaac Newton's magnificent natural philosophy, which used those very mechanical principles to describe the world as a "mighty machine."[1] People even began to refer to Newton's natural philosophy as a "mechanical philosophy," an early form of philosophy of technology.

New technology was thus an integral feature of the Scientific Revolution. An interesting and often overlooked fact is that technology flourished during the Middle Ages. A great many striking inventions saw the light of day, not only gunpowder and printing but the compass, new kinds of ploughs and water wheels, the spinning wheel, and other machines. Yet all these inventions, which were to radically change the world, remained narrowly confined to professional artisans and were largely ignored by the leading philosophers of the time. An exception was Francis Bacon (1561–1626), whose many roles included that of active philosopher. Technology was a particular interest for Bacon, and his claim that knowledge is power implies that human beings, through their knowledge of nature and their exercising of technological power over nature, can create a rich and prosperous society—an ideal society.

Apart from Bacon's optimistic views on the boundless possibilities of technology, leading philosophers of the seventeenth and eighteenth centuries had almost little of significance to say about technology. Technology was generally regarded as applied science and as a force entirely for good.[2]

Not until the Industrial Revolution of the late eighteenth century did philosophers begin to ask critical questions about technology. The notion that technology might have a negative impact on human life first emerged in the Romantic movement. Romanticism arose in part as a reaction to the Industrial Revolution, which created ugly and polluted industrial towns populated by an urban lumpenproletariat. For Romanticism, the Industrial Revolution was closely bound up with a new technology that destroyed both nature and people. For example, in his essay *Discours sur les Sciences et les Arts* (1750) Jean-Jacques Rousseau took issue with his contemporaries by criticizing the Enlightenment idea that science and technology automatically furthered the development of society. On the contrary, he argued, the emergence of modern science and technology had been directly harmful to society and morals.

There followed a rich and varied stream of Romantic critiques of modern technology and science for having destroyed people's lives. For instance, Romantic thinking informs the work of the American writer and philosopher Henry David Thoreau (1817–1862), whose ecologically inclined book *Walden* (1855) advocated a simple life in nature over the inhuman environments of the burgeoning cities and their mass-production industries. A later example is the historian Oswald Spengler (1880–1936), whose magnum opus, *Der Untergang des Abendlandes* (1918–22),

[1] Carl Mitcham, *Thinking through Technology*, p. 19.

[2] Dusek och Scharff, *Philosophy of Technology*, p. 1.

describes a thousand-year-old civilization undergoing a crisis akin to the way that many people experience the approach of their twilight years. In a few hundred years, Spengler argued, our culture would be as "dead" as classical antiquity had been by the time of the Germanic Invasions.

In Spengler's short book *Der Mensch und die Technik* (1931), which can be seen as a sequel to *Der Untergang des Abendlandes*, Spengler discusses technology, which he regards as central to Western civilization. Spengler defines technology as "the tactics of living":

> *Technics is not to be understood in terms of the implement.* What matters is not how one fashions things, *but what one does with them*; not the weapon, but the battle. This battle *is* life—life, indeed, in the Nietzschean sense, a grim, pitiless, no-quarter battle of the Will-to-Power.[3]

Spengler here wishes to warn of the consequences of humanity's blind worship of this technology. It is natural but also irresistible, he argues, and it rules over us like a master rules over a slave. Just like the Romantics, Spengler is reacting to machine society and the great industrial cities and advocating their replacement with simpler ways of life that are more in harmony with nature.[4]

The technological optimism of the Enlightenment forms a sharp contrast to the bitter criticisms of technology made by Rousseau and the Romantic movement. Philosophers such as Diderot and Voltaire argued that society should be governed by Reason in the form of an "enlightened despot" of high moral and intellectual stature. In the early 1800s, this vision was given impetus in differing ways by Hegel and Comte. Comte, who was technically trained, rejected Hegel's "speculative metaphysics," and in his *Cours de politique positive* he instead chose to expand on the somewhat sketchy notions of his teacher, Henri de Saint-Simon, by describing a "technocratic utopia" based on technological-scientific (or "positive") principles. Comte's philosophy, *positivism*, has had an extraordinarily profound influence on modern science, particularly social science or "sociology," of which Comte was also the founder. For his part, Hegel described how of Reason or Spirit becomes self-realized in the world in works such as *Phänomenologie des Geistes* (1807) and *Grundlinien der Philosophie des Rechts* (1821), which were to be of fundamental importance to the emergent philosophy of technology.

Philosophy of technology as we know it is a relatively new area within philosophy. Only in the mid-nineteenth century did philosophers begin publishing works devoted exclusively to the topic. Questions about and for technology thereafter became an increasingly frequent object of abstract philosophical reflections, which thereby acquired a new target and a new direction. Those who became involved in this new branch of philosophy were primarily neo-Hegelians, such as Ernst Kapp and Karl Marx, but by the turn of the twentieth century neo-Kantians and other

---

[3] Spengler, *Der Mensch und die Technik,* pp. 8–12.

[4] Many philosophers have examined the problematic aspects of modern technological society, including Nikolaj Berdjajev (1874–1948), Karl Jaspers (1883–1969), and Gabriel Marcel (1889–1973).

technologically inclined philosophers were also beginning to ask questions about technology.[5] Their starting point was generally technology itself, and they typically accepted technology and technological development or took them as givens. Many of them, particularly the Left Hegelians, had a tendency of emphasizing the material aspects of technology, and they sought to counter the possible negative consequences of technology by applying ethical and political correctives.

The first person to coin the expression "Philosophy of Technology"(1877) was the German philosopher Ernst Kapp (1808–1896), who, like his younger contemporary Karl Marx, was a committed Left Hegelian.[6] Like Marx, he sought to translate Hegel's dynamic idealism into a more down-to-earth materialism. But where Marx's materialism sought to apply Hegel's philosophy of history to the new science of economics, Kapp applied history to geography. Published while he was still young, his geographical study *Vergleichende allgemeine Erdkunde* (1845) anticipated what we now call environmental philosophy. In it, Kapp shows how geography and environment leave their mark on the socio-cultural order. In terms that could have been taken from Montesquieu's theory of climate, he describes how untamed nature affects not only economics and culture but also political structures and military organizations. On the other hand, Kapp's special version of the Hegelian dialectic insisted upon the gradual "colonization" and transformation of this inhospitable environment.

Drawing an analogy with Hegel's description of the master-slave relationship in *Phänomenologie des Geistes*, Kapp argues that people must try to meet the challenges that their inhospitable environment presents—they must try to tame the wildness of nature. This requires a "colonization" of both space (via farming, mining, architecture, road-building, etc.) and time (via telegraphy, etc.). Kapp imagines a global telegraphy system that connects all of the world's countries, peoples, and languages and that makes Earth habitable.

In his most important publication, *Grundlinien einer Philosophie der Technik* (1877), Kapp offers a detailed and systematic description of how instruments, tools, and weapons should be understood as different kinds of prostheses or "organ projections." For him, the term "organ projection" refers to the fact that a particular organ in the human body creates an exterior counterpart to itself. Every instrument,

---

[5] One philosopher who took an interest in technology was the Russian engineer Peter K. Engelmeyer (1855-c.1941). His four-volume *Filosofia tekhniki* (Philosophy of Technology, 1912) is the first general overview of problems in philosophy of technology. Engelmeyer was the first to systematically connect technology to politics and other philosophical topics. In France, examples of technological oriented philosophers include the social scientist *Alfred Espinas* (1844–1922), the engineer *Jacques Lafitte* (1884–1966), and, more recently, the psychologist and philosopher *Gilbert Simondon* (1923–1989). In the Netherlands, this tradition includes the engineer *Hendrik van Riessen* and, in the rest of the world, the Spanish-Venezuelan *Juan David García Bacca* (1901–1992) and the Argentinian-Canadian philosopher *Mario Bunge* (b.1919). The latter seeks to explain reality in technological-scientific terms and to reformulate the task of humanistic thinking (including philosophy itself) in a technological-scientific direction.

[6] My discussion here on Kapp, as well as on Dessauer, Mumford, Ortega y Gasset and Ellul further on is mainly based on and inspired by Carl Micham's classical *Thinking through Technology*.

tool, or weapon is thus an externalization of our hands. Hammers, axes, chisels, drills, knives, and revolvers are all examples of projections of our hands. In *Grundlinien*, Kapp also draws a parallel between external technology and what is inside a human being: for example, he describes the railway as an externalization of the circulatory system and the telegraph as an extension of the nervous system. Kapp also treats non-material entities such as language and the state as extensions of mental life.

Many of Kapp's fundamental ideas were shared by Karl Marx. Although, as has already been noted, both were Left Hegelians, Kapp sought apply Hegel's philosophy of history to geography, while Marx sought to connect it to the new *theories of economics* and the fledgling *socialism* that had begun to emerge, above all in France.[7]

Capital, the first volume of which appeared in 1867, contains the kernel of Marx's philosophy of technology in Chap. 12, "The Division of Labour and Manufacturing," and Chap. 13, "Machinery and Large-scale Industry." He describes how the tools of production evolve from simple utensils to advance machines and how these machines are organized into complicated systems in which the workers, like Charlie Chaplin in *Modern Times*, become integrated components. Workers are alienated by being deprived of any control over the production process, the tools of production, and the fruits of their labour, ultimately being reduced to mere cogs in the machinery of capitalism. And workers under capitalism are not deprived solely of their labour but also of their very nature, their humanity. The precondition for this transformation, Marx argues, is "modern technological science," which analyses the production process down to the minutest detail. Marx shows how this new technology undermines traditional craft skill and pride by removing the worker to a ghostly, automatized factory in which the various stages of production are homogenized and interchangeable.

At the same time, Marx is surprisingly modern in his admiration for mechanical technology, and, just like Kapp, he shows respect for technology's capacity to tame the unfettered forces of nature. The goal of production is admittedly to satisfy an immediate need for consumer goods, but its higher purpose is in fact production itself, which is the precondition for everything. The essence of human beings lies not in themselves but in their *productivity*. They are primarily defined not by thought but by action. They are by nature active, creative beings whose ultimate goal is to change inhospitable nature into a human world. In other words, whereas technology under capitalism is currently a vehicle for exploitation and the oppression of workers, it will be a force for good in the communist society that lies in the near future. According to Marx, modern technology will achieve that which philosophy and religion could not: it will restore human freedom.[8]

---

[7] For a good overview of Marx's philosophy, see *The Stanford Encyclopedia of Philosophy.*

[8] Marx's analyses were subsequently developed in ways that warrant closer examination. The Marxian legacy is strongest in the work of Max Horkheimer, Theodore Adorno, Herbert Marcuse, and Jürgen Habermas. All belonged to the neo-Marxist Frankfurt School, and in their "critical theory" a critique of the shortcomings of economics evolved into a questioning of the Enlightenment,

In contrast to Kapp's and Marx's "materialist" view of technology, the neo-Kantians sought to ask questions about technology's purpose and legitimacy on the basis of the "idea" that it embodies. "Technology's origin lies in ideas," argued their leading representative, Friedrich Dessauer (1881–1963).[9] In his most important work, *Philosophie der Technik* (1927), Dessauer offers both a Kantian interpretation of the transcendental preconditions for *force* in technological-scientific knowledge, and reflections on the ethical consequences of its application.

Dessauer proposes that the three Kantian critiques—the Critique of Pure Reason, the Critique of Practical Reason, and the Critique of Judgement—be supplemented by a fourth: a critique of technological creation. In the process of technological creation, argues Dessauer, a contact is established with the thing-in-itself. The essence of technology lies not in the industrial production process or in its end products (which are used only by the consumers) but in the *technological creation.* Dessauer refers to an "inner processing" that puts consciousness in contact with a "fourth world," a world of "preestablished solutions to technical problems." This inner processing is what connects us to the transcendental thing-in-itself. Technological creation brings forth "existence out of essence," the material embodiment of transcendental reality.[10]

Although philosophers often regard Dessauer's reworking of Kant as naïve and simplistic, it can be thought of as a kind of extension of the Kantian project. For Kant, the transcendental is bound up with moral and aesthetic experience. Dessauer nonetheless connected the transcendental thing-in-itself with a kind of moral experience that, he argued, Kant refused to acknowledge—modern technology.

---

together with a realization that the problem of technology was not restricted to economics but applied to culture as a whole. The problem with the Enlightenment was that it replaced religion with "instrumental reason" without offering an objective theory of how that reason should be used. The result was that the Enlightenment became turned against itself, with production in society becoming directed towards the military and the entertainment industry. Horkheimer and Adorno were highly critical of electronic media, such as television, which they saw as inducing passivity and stupidity. Unfortunately, they were unable to move beyond this critique to formulate a working theory of reason and the correct use of technology. Marcuse wanted to devise specific protocols for technological power. To this end, he sought to transfer technology and scientific thinking to marginalized groups in society and other counter-cultures of artists, poets, environmentalists, feminists, and others. He wished to "liberate" traditional science and technology, which he saw as corrupted by the logic of power, and replace it with an alternative form of science and power that started from human values. For his part, Jürgen Habermas aligned himself with the more Marxist view that the problem lies not in science and technology but in the social conditions under which they exist. In *The Theory of Communicative Action* (1981), he formulates a theory of "communicative action" intended to serve as a concrete guide for technological and political development. The problem, argues Habermas, is not instrumental reason—which is indispensable in technological-scientific contexts—but the fact that it has been allowed to invade our lifeworld, including political life and private life. Habermas speaks of technocracy's attempts to "colonize the life-world," by which he means the deployment of a technocratic approach and instrumental reason within the sphere of human communication.

[9] Dessauer, *Philosophie der Technik,* p. 146.

[10] Dessauer, *Streit um die Technik*, p. 234.

For Dessauer, technology thereby acquires a moral and, indeed, mystical, significance. Technology is not limited to the benefits it brings but is primarily characterized by a Kantian imperative or divine vocation: modern technology "participates in creation … it is the greatest experience for us mortals on earth."[11] For Dessauer, technology is a religious experience—and religious experience has a technological significance.

Although posterity has shown little interest in Dessauer's somewhat idiosyncratic line of argument, his wide-ranging, interdisciplinary philosophy of technology and his theory of "*homo inventor*" have nonetheless had a considerable influence well into our own period.[12]

The First World War marked the emergence of technology as an important philosophical question, and in the 1920s it became a focus of general interest, not only among intellectuals. By 1930, every aspect of human life had been impacted by the rapid advent of new technologies—in the home, at work, and in society at large. Animals were replaced as means of transport, electricity came into general usage, as did the telephone, the gramophone, and, in particular, the radio, which instantly brought far-distant regions of the world into people's homes. The architectural historian Siegfried Gideon aptly called the decade between 1920 and 1930 "the age of total mechanization," a period when every moment in daily life was transformed by a legion of technological innovations.[13] Another prominent commentator of this period, Ernst Jünger, used the concept of "total mobilization" to describe the era as

---

[11] Dessauer, *Philosophie der Technik,* p. 66.

[12] An important thinker of the same generation as Dessauer was Ernst Cassirer, whose suggestive and thought-provoking essay "Form and Technology" (1930) seeks to answer the question of "technology's being." According to Cassirer, technology must be understood as an *activity* and not as a "dead product," a kind of production or intellectual creativity that is intimately connected to thought. Technology makes possible thought as such by providing the basis for "a kind of intermediary belonging to the essence of thinking" ("Form and Technology," p.30). As important as Cassirer's positive view of productive human activity as an essential factor in every kind of knowledge production is his conviction that knowledge can only be acquired by means of a combination of historically constituted symbolic forms and external material assistance. In a polemic against Kapp, he also argues that the technologies of the new era are not only extensions of the body, i.e. organ projections, and that technological artifacts are no longer modelled on nature. Flight, Cassirer says, became possible for human beings only after they created fixed wings, which have no counterpart in nature (ibid., p. 54). Alongside technologies that simply imitate nature, there are, in other words, "purely symbolic technologies"—that we today call information technologies—that have no basis in analogy or imitation.

Interest in Cassirer's philosophy of technology is currently burgeoning, with new books and translations being published continually. An example is the anthology *Ernst Cassirer on Form and Technology* (2012), which contains a new translation of Cassirer's "Form and Technology" together with contemporary readings of Cassirer's writings, such as Hans Ruin's essay "Technology as Destiny in Cassirer and Heidegger: Continuing the Davos Debate," which examines "Form and Technology" alongside Heidegger's "The Age of the World Picture" and "The Question Concerning Technology." Finally, it should be mentioned that Derrida was familiar with Cassirer's philosophy and cites him approvingly on several occasions (see, for example, *De la grammatologie*, p. 156–7 and *Margue—de la philosophie*, p. 224).

[13] Krois, *The Age of Complete Mechanization*, p. 55.

a way of showing that the military operations of the First World War were no longer confined to armies fighting battles far removed from everyday life; rather, war was a life-and-death battle for all society and for every country involved.

Ernst Jünger himself served as a soldier in the First World War, was badly wounded several times, and received a number of decorations for conspicuous valour on the battlefield. In his essay *Die totale Mobilmachung* (1931), Jünger takes his starting point in his personal experiences of the new, advanced military technology and the colossal material destruction on the battlefields of the First World War. He coined the notion of total mobilization as an attempt to describe the era of modern technology as such. Jünger paints a deeply pessimistic picture of his time as a world in which "the new human" takes on the role not only of technology's master but also its slave. Perhaps his most important work was *Der Arbeiter*, which he completed at the end of the decade and published in 1932. A major inspiration to Heidegger's later philosophy of technology, it can be seen as a political manifesto of modernity. Its subject is the new kind of human exemplified by the worker, the Industrial Revolution, and the relationship between the worker and technology. In one of its central chapters, which deals with technology and technicity, Jünger defines technology as "the mobilization of the world in the image of the worker."[14] No longer an aid for improving humanity's conditions of life, technology is here the language of a world at work, a general language that follows its own laws beyond human control. Individual technological implements are as subordinated as human beings themselves: both are simply means by which this technology expresses itself. According to Jünger, this technological world and its technologized workers were the only things to survive the destruction of the First World War and the despair of the postwar years.

Jünger's doctrine of technology revives conservative and collectivist values that have their roots in particular classical ideas and that contrast sharply with the Enlightenment's individualistic and universalist view of humanity. The nineteenth century's faith in progress is doomed to be replaced by this technological world defined by total mobilization. Its understanding of technology as an agent of progress is merely one face of a Janus-headed technology, the other being its military potential. With fatal prescience, Jünger's book anticipates the total mobilization of fascism, Nazism, and Soviet communism in the face of impending catastrophe.

Although other thinkers such as Oswald Spengler warned of the outbreak of a high-tech war, it was only with the use of atomic bombs against Hiroshima and Nagasaki at the end of the Second World War, and the realization that nuclear weapons threatened to destroy all humanity, that the demand for a critical evaluation of technology became a pressing item on the agenda. During the 1960s and 1970s, this urgency was reinforced by a view that technology was fundamentally military, industrial, and large-scale in nature as well as by a fast-growing global

[14] Jünger, *Der Arbeiter*, pp. 149–94.

environmental awareness and, not least, by public concerns about genetic technology and the possibility of human cloning and the manipulation of human nature.[15]

Partly as a response to this demand, the Society for Philosophy and Technology was formed in 1976. SPT is a world-wide organization for philosophy of technology in which philosophers with different backgrounds and approaches meet for "cross-border" discussions of philosophical-technology issues. There is, however, not complete consensus as to what is meant by the designation philosophy of technology. In large part this is because of the close connections between a number of other philosophical fields, such as the philosophy of science, ethics, aesthetics, and metaphysics as well as political philosophy and philosophy of religion. Working with philosophy of technology therefore requires philosophical instincts that are both deep and broad. We might echo the question posed by French philosopher of technology Jacques Ellul: how can humanity be expected to exercise control over technology when an overview of all these philosophical fields is impossible?

To begin with, we have to ask what we mean by technology. One answer may be that technology is, in Carl Mitcham's definition, the "making and using of artifacts," that is to say, a deeply practical activity.[16] However, a number of philosophers do not regard it as meaningful to propose definitions or general characteristics of the concept. Some of them, particularly those scholars with an empirical focus, take issue with Heidegger and Ellul by arguing that there is no "essence of technology" and thus no possibility of a general definition of technology.

The science-technology relation holds a particular interest for philosophy of technology. Today, there is a widespread view that science and technology are becoming increasingly intertwined. Because science has now merged with and become dependent upon advanced technological apparatuses, technology itself has become the precondition and driving force of science. This notion runs directly contrary to the view of technology as "applied science" in which science precedes technology and drives it forward. The concept "techno-science" precisely defines technology and science as inextricably interwoven. The idea that technology philosophically precedes science has much in common with Heidegger's later view of technology as the fundamental force driving modernity.

The humanity-technology relation is likewise a central question in philosophy of technology. There have been many attempts to define what a human being is. While some have emphasized the making of tools, others deny that this kind of activity is

[15] In my view, the public's critical view of technology shifted after 1989. After the fall of the Berlin Wall and the end of the Cold War, a new political and cultural situation arose, a new climate in which military knowledge was converted into civilian knowledge, for example, into environmental knowledge through the use of geostationary satellites to measure global natural resources, temperature changes, changes in sea level, and the like, knowledge that had previously only been available to the military. This more positive approach to technology was reinforced by the advent of new communication technologies and media such as the World Wide Web, email, and mobile telephones, which appealed particularly strongly to children and young people. There are perhaps as many naïve technophiles today as there were technophobes during the previous period.

[16] Mitcham, *Thinking through Technology*, p. 1. For his "definition" of the concept "artifact," see pp. 161–5.

a central feature of human nature; rather, the most important human attributes are consciousness, language, or the use of symbols. How scholars see the matter is largely dependent on their particular attitude towards technology. Those who regard technology as largely beneficial tend to regard humans as tool creators. Those who see technology as a danger or a threat tend to emphasize consciousness (previous philosophers) or language (contemporary philosophers).[17]

Ultimately, the debate is about whether humans first acquired consciousness, then stood up and, their hands now free, began making tools—or whether they stood up first and made tools, and then acquired consciousness.[18] A third alternative would perhaps be to regard the emergence of humans as a gradual process in which humans and technology are essentially two elements in the same dynamic feedback system.

Following this introductory survey of the historical background to philosophical technology, I will proceed to a discussion of current philosophy of technology. First, I address the phenomenologically and transcendentally oriented philosophy of technology that, in differing ways, describes technology as an all-encompassing, autonomous, and alienating force. In the subsequent section, I discuss postphenomenology and empirically focussed philosophy of technology, which can be seen as a development of the former.

2.2 Phenomenology and Transcendental Philosophy of Technology

How does phenomenological analysis view technology and our relation to technology? Although phenomenologically oriented philosophers jointly use certain concepts, such as intentionality, phenomenon, and lifeworld, their philosophical positions are often very different. Even so, they share a fundamental perspective with a range of other philosophers: for them, technology is not only an all-encompassing, autonomous, and alienating force but something that exists beyond individual things, a kind of network that constitutes the condition of possibility for knowledge and meaning. In order to do justice to the various positions accommodated within this framework, I will examine in this section six ways of approaching the question of technology that have been, and continue to be, of major importance for modern philosophy of technology.

---

[17] Dusek och Scharff, *Philosophy of Technology*, pp. 112–3.

[18] Ibid., p. 117.

2.2.1 Phenomenology and Lifeworld

Edmund Husserl, the founder of modern phenomenology, may not have written a separate study of philosophy of technology, but his ideas have been, and continues to be, enormously important to it. Philosophers of technology of various persuasions have been deeply influenced by Husserl's phenomenological thought, such that it is barely an exaggeration to say that a considerable number of contemporary philosophers of technology have their intellectual roots, in one way or another, in Husserl's phenomenology.

It was in his *Logische Untersuchungen* (1900) that Husserl first formulated his goals and theories. To simplify somewhat, phenomenology can be described as an attempt to bracket all preconceived notions about the outer world. Husserl retains only the "pure" knowledge that occurs in acts of consciousness. The task of phenomenology is to provide an adequate description of the experience that is revealed in these theoretically pure acts of knowledge. That which reveals itself is what Husserl calls *phenomenon*, while description of such phenomena is what he calls *phenomenology*.

It is impossible to ignore the strong element of idealism in Husserl's phenomenology, in which the pure, ideal content of acts of consciousness constitutes true reality. Philosophers should, like Plato, look into the world of ideas and devote themselves to the intuition of essences (*Wesenschau*). By contrast, the actual world, the outer world, which Husserl brackets, no longer seems essential.

In his final and incomplete work *Die Krisis der europäischen Wissenschaften und die transcendentale Phänomenologie* ("The Crisis of the European Sciences and Transcendental Phenomenology," part of which appeared in 1935), Husserl tries to give phenomenology an ultimate foundation. Although the ideal and timeless world that Husserl had previously sought to describe through observing essence rejects every kind of historicism, we as humans live our daily lives *in time*. With the passing of the years, Husserl himself became increasingly aware of this problem in phenomenology, and he therefore sought in his final work to attribute a kind of historically determined situatedness to phenomenology.

In *Krisis*, Husserl notes that our theoretical way of looking at the world can be traced back to an earlier tradition. At the same time, he argues that the very idea of being able to experience the world theoretically is in itself a tradition whose origins are generally associated with classical Greece. But if the Greeks' "theoretical attitude" really is new, it needs to be differentiated from an originary attitude towards life based on some originary and naïve lifestyle. Husserl here refers to an "original natural attitude" and the "original natural life." This naïve "pre-theoretical" life is lived in a world that is the horizon of its activity. The world within which the scientist or philosopher is concerned is precisely the world-horizon within which this naïve life takes place—*lifeworld* (*Lebenswelt*). Lifeworld is always "already there" or "pre-given" (*Vorgegeben*) whenever theory begins its work. What surrounds us is this directly apprehended world, coloured by our expectations, feelings, values, and needs. Lifeworld is an immediate sensory world but also an interpersonal world of social structures and conventions. Lifeworld is the practical life that surrounds theoretical life; it is "pre-theoretical" and, as such, constitutes a condition of all rational, scientifically informed activity. Ever since Galileo, as Husserl sees it, scientific thinking has insufficiently addressed this precondition in its quest for scientific truth.

Europe's spiritual crisis—the anti-rationalism that Husserl sees around him—actually involves a threat to the European idea of rational, scientific progress. In an effort to address the problem, Husserl sought to develop a theory of rational progress by showing how eidetic fields of knowledge—of which mathematics and geometry are prime examples—were able to emerge from more empirical and material intellectual environments.

The crisis of modern science, which forms the central theme in this second part of Husserl's *Krisis*, is that scientific thinking is no longer capable of reflecting upon its own premises. For Husserl, this crisis is a direct consequence of the rampant *objectivism* that has predominated ever since the Renaissance's scientific revolution. The crisis is thus not a crisis in science itself but a crisis that relates to the "significance of the sciences for life," human life.[19] Because this is deeply existential crisis, Husserl is able to write:

> In our vital need—so we are told—this science has nothing to say to us. It excludes in principle precisely the questions which man, given over in our unhappy times to the most portentous upheavals, find the most burning: questions of the meaning or meaninglessness of the whole of this human existence. … In the final analysis, they concern man as a free, self-determining being in his behavior towards the human and extrahuman surrounding world *(umwelt)* and free in regards to his capacities for rationally shaping himself and his surrounding world.[20]

According to Husserl, modern science has made two fundamentally important advances: firstly, generalization of Euclidian geometry as a formal *mathesis universalis*; and secondly, mathematization of nature. The first innovation builds on classical scientific knowledge but supersedes it in the same way as the infinite supersedes the finite. The fundamental novelty here lies in the idea of an infinite totality that can be quantified by means of a generalized mathematics. By means of this innovation the realm of *absolute exactitude* has now been conquered. The second innovation is associated with Galileo, the scientist who creates an entirely new natural science in which nature is approached mathematically in the same way as ideal geometric figures. Yet Galileo is not satisfied with mathematizing the quantitative side of nature but further claims that its material qualities are purely "subjective" illusions that can also be approached mathematically. In fact, Galileo argues that the need to approach all of nature mathematically "follows as a matter of course." For Husserl, Galileo's mathematization of nature involves a "technification" of thought in the sense that scientific thinking becomes an art, namely the art of using fixed technological rules to calculate that which can actually only be verified through practical experience and things themselves. Science becomes a manipulation of abstract

---

[19] Husserl, *The Crisis* § 2 is headed: "The crisis of science as the loss of its meaning for life."

[20] Husserl, p. 6.

symbols much like chess or cards, whereby *originary* practical thinking, that which gives meaning to this technological process, is forgotten. Science is thereby alienated, and consciousness has lost the key to itself.

Husserl's critique of science's "technification" is nonetheless not a critique of scientific objectivity. On the contrary, for Husserl, it is the ultimate and highest form of knowledge. Science's ideal objectivity is closely associated with *writing* and written expression. It is not simply that scientific meaning becomes definitively objective by being written down. As written down, meaning can also be safely passed on to later generations and revived, forming a joint reserve of experience that generations of scientists can share in and build on. In "The Origin of Geometry," Husserl writes that the huge and complex theories that have been developed down the centuries would not have been possible without writing's documentary and preserving function.[21]

It is important to note that Husserl's treatment of "writing" in "The Origin of Geometry" is Derrida's own source of inspiration when he develops the concept of "arche-writing." And, as we will see in the next chapter, the question that Husserl asks about geometry's originary meaning is central not only to Derrida's development of "arche-writing" but also for his understanding of technology.

2.2.2 *Existential Phenomenology*

Before long, Husserl's phenomenology was being developed in several different directions. One of these focussed on the question of how technology affects human existence. This question was posed explicitly by the existentially oriented German philosopher Karl Jaspers, whose writings discuss technology in great detail. Another was José Ortega y Gasset, possibly the leading philosopher of technology in the Spanish-speaking world.

Ortega spent the years just prior to the First World War in Germany, where he studied Husserl's phenomenology. In his first book, *Meditaciones del Quijote* (Meditations on Quixote, 1914), he outlined a new theory of the human in which he criticizes Husserl's analysis of consciousness. Instead, he refers to an "existential intentionality" or a "concrete human life" that co-exists between the self and "its circumstances." Ortega coined the phrase "I am I plus my circumstances" in reference to a subjectively interpreted outer world, a situation—a line of thinking that bears some resemblance to Heidegger's in *Being and Time*.

Ortega championed existentialism long before the term (which he avoids) even existed. He instead preferred the expression "ratio-vitalism," and later he also referred to reason as "historical reason." For Ortega, however, reason is not disembodied and impersonal but something with which we are equipped in order to be able to manage concrete issues in life or historical situations.

---

[21] Husserl, "The Origin of Geometry," pp. 183–189.

Twenty-five years later, he published *Meditación de la técnica* (Meditation on Technology, 1939), a short study whose title indicates the importance that he ascribed to technology. The book opens with a discussion of the relationship between humanity and technology. According to Ortega, technology is necessarily a fundamental aspect of human life. Ortega's philosophy of technology starts from the idea that human life exists in relation to "circumstances." Unlike a stone, a tree, or an animal, human existence is not permanently fixed but is an interaction between the idealism of the self and the empirical reality of its actual surroundings. Human nature is something that humans must continually create for themselves. Human beings—as Sartre says—are their own project, but, adds Ortega, always beyond themselves.

According to Ortega, this self-creative or self-interpretative undertaking has two stages. In the first, human beings create a notion of the life project that they want to fulfill (inner innovation), while the second stage involves the material realization of this project (external innovation). The technological requirements will depend on which project the individual wants to realize, which means that there will be as many forms of technology as there are life-projects.

From one perspective, it could be claimed that Ortega is defining human beings as *Homo faber*, assuming that *faber* is not limited to material production but also includes spiritual or cultural creation, for example, of art and literature. Technology can also be seen in this context as a "human organ projection," albeit not in Kapp's material sense.

In the modern era, discovering the technological means to achieve *any goal whatsoever* has become a scientific method or technology in itself. Technological science has thereby attained methodological perfection. This technicity of modern technology involves a new way for consciousness to operate, which eventually leads to a huge problem for modernity: the imagination, our originary creative capacity whose main task is to formulate human ideals, threatens to wither and disappear. Human beings will gradually lose the capacity to choose anything, even as technology becomes in the hands of hardheaded technicians "an empty form—like the most formalistic logic; it is unable to determine the content of life."[22]

2.2.3 *Hermeneutic Phenomenology*

If Ortega represents the alliance of phenomenology and existential philosophy, then Martin Heidegger is the threshold figure for a hermeneutically oriented phenomenology that emphasizes the philosophical significance of interpretation and meaning. From a hermeneutic perspective, the key question is not how humanity can realize itself in the world but how the world appears to humanity. Heidegger's hermeneutic perspective has, and has had, an extraordinarily profound importance for

[22] José Ortega y Gasset, Meditación de la técnica, p. 366.

modern thinking on technology. Indeed, he can be described as the founder of modern philosophy of technology. Even today, his reflections on technology exert a very powerful influence on debates within philosophy of technology. However, while they exhibit some similarities with Ortega's thinking, these reflections are not easy to summarize.

In discussing Heidegger's relation to technology, two things must be borne in mind. First, Heidegger is part of a long tradition, going all the way back to antiquity, in which the task of the philosopher is to ask questions rather than providing answers. Second, the overarching issue in Heidegger's philosophy is the question of Being. For Heidegger, the question of technology is closely bound up with the question of Being.

In *Being and Time* (1927), Heidegger describes human existence's way of being-in-the-world. We humans do not exist primarily as observers or as theorizing beings. We are occupied with practical concerns and problems. For this reason, the world chiefly presents itself to us as something "ready-to-hand" rather that something that is simply "present-at-hand." What is "ready-to-hand" are the instruments or tools that we use in our daily activities. The opposite to this being "ready-to-hand" is being "present-at-hand," which is the "scientific" view of these instruments or tools, namely as something we observe or think of as independent of and outside of ourselves. Heidegger's famous example is the hammer. When we use a hammer to hit a nail, we are entirely occupied with practicality—driving a nail into wood. But as soon as the hammer breaks, our interest switches to the theoretical—the hammer itself.

In *Being and Time*, Heidegger therefore understands technology in terms of useful tings that he calls "equipment," *Zeug*. Equipment opens up a world to humans. When someone uses equipment, a world appears that encompasses not only the manufactured object but also the material in which it is produced, its future user, and the setting in which it will be put. Equipment is the central point around which a world is created.

A slightly different view of technology emerges in Heidegger's essay "The Origin of the Work of Art," one that represents a cautious step towards what would become his famous "turn." In addition to equipment, he refers here to another kind of object that can also open up a world: the work of art. A work of art "is" in a way that is very different from equipment. In contrast to equipment, which only appears when it is being used, the artwork reposes in itself. In the independence of the work of art, Heidegger argues, being can appear in order to create a world. For example, when we look at a painting by Van Gogh depicting a pair of shoes, we are "suddenly somewhere else than we usually tend to be."[23] The painting allows the shoes to appear, it un-veils them, brings them into unconcealment. The work of art therefore belongs to the domain of *alētheia*, a Greek word that Heidegger translates as "truth." The work of art should not only be thought of in terms of beauty or as a representation of reality but must primarily be understood as "the truth of Being." What the

[23] Heidegger, *The Origin of the Work of Art*, p. 35.

work of art does is allow Being to appear, to be un-veiled: "The artwork opens up in its own way the Being of beings."[24]

Heidegger goes on to discuss the bringing forth of a work of art as a "creative act," while he refers to the bringing forth of equipment as "making "(*Anfertigung*). But what distinguishes "creation" from "making," he asks himself? To answer this question, he goes back to the classical Greeks, who called "making" *techné*, a word denoting both "handicraft" and "artwork." But the Greeks regarded *techné* not as a form of practical activity but rather as a mode of knowing, a *seeing* in the widest sense of the word. Seeing here means perceiving what is present as such, making Being emerge from its concealment. *Techné* as a mode of knowing is a form of unveiling that brings forth the originary truth of Being, *alētheia*. It is precisely this bringing forth of the truth that takes place in all art and in all artistic activity, that is, not only in the work of art *itself* but also in the *creating* of it.

This original meaning of *techné* (technology) was regarded by the ancient Greeks as the opposite of *physis* (nature). The technological was seen as a kind of artificial substitute for nature, a prosthesis that imitated and perfected *physis*.

After Heidegger's so-called "turn" in the early 1930s, he became increasingly interested in the question of the steadily increasing technologization of our civilization. For Heidegger, the advance of modern technology brings with it a new view of humanity and its relation to the world. Max Weber has described this new view as a "disenchantment" of the world, with the world's mystery being replaced by techno-scientific rationality. Technology as the "disenchantment of the world" necessarily leads to a loss of faith in religion and, with it, an ultimate meaning to the world. The question of technology is thus intimately connected with nihilism.

Heidegger is one of the few philosophers to have *fundamentally* engaged with the great question of technology and what technology really *is*. In fact, the question of technology is a central theme in particular of Heidegger's later writing from the 1950s and after. With increasing intensity, Heidegger asks questions about modern technology and its associated modern lifestyle defined by rampant consumption, superficial values, and the ruthless exploitation of nature. Although Heidegger's thinking unquestionably contains an element of criticism of technology, it also contains a conviction that the era of modern technology must run its course before something new and better can arise.

Even if Heidegger discusses question of a technological nature relatively often, what matters most to him is *not* correctly describing technology itself. Instead, he wants to engage with the question by reflecting upon the essence of technology. In one of his best known and most influential essays, *Die Frage nach der Technik* from 1954 (*The Question Concerning Technology*, 1977), he criticizes the prevailing view that technology is primarily instrumental, that is to say, a neutral aid or a human activity. Instead, he claims that "the essence of technology is by no means anything technological."[25] This fundamental statement is at least partly a

[24] Ibid., p. 38.

[25] Heidegger, *The Question Concerning Technology*, p. 4.

philosophical statement since it makes it possible for philosophical thinking to ask questions about the essence of technology from a position located outside of technology itself. Only on the basis of such a position can we understand what technology *is* and formulate a satisfactory answer to the problems that modern technology has created and continues to create.

However, Heidegger also held that the original meaning of *techné* had been lost. In his view, this loss was the result of a continuous historical process that begins with Plato's devaluation of art and nature as well as Aristotle's (and later Descartes's) view of knowledge as a way of mastering nature. With the advent of modernity, *techné* begins to be understood in a limited sense as "technology." According to Heidegger, our own understanding of modern technology has come to be associated with the manipulation of nature through the application of technological-scientific knowledge for industrial production. Modern technology has thereby unveiled itself in a very special way: as a forgetting of the truth of Being. The latter is now merely a distant echo of its true origins. For modern humans, the world is nothing more than a resource for technological manipulation within the framework of its usefulness. This framework or "enframing" (*Ge-Stell*) is the way in which the whole meaning of the world emerges. For Heidegger, "enframing" is the *essence* of modern technology.

In our technological culture, Heidegger argues, it is no longer humans who control technology but rather technology itself that appears as a defining frame or enframing within which everything is interpreted. Human beings are themselves "enframed" within technology. Enframing is the source of the underlying attitude of modern people, not forming the basis for modern technological development but also permeating the way we view practically everything in the world. More or less everything is treated as "resources" to be organized and used. The world as a whole is primarily viewed as a vast reservoir of potential products that can be extracted and manipulated in order to serve the goals and desires of human beings. Consequently, enframing—as the essence of technology—becomes synonymous with *Being itself*. Human beings as "enframed" now appear as the manipulator and exploiter of others, or, vice versa, as the helpless victims of manipulation and exploitation. Their entire "life" becomes *totally mobilized*.[26]

As noted earlier, the concept of "total mobilization" was coined by Ernst Jünger. Heidegger began reading Jünger in the early 1930s, and Jünger's writings, particularly "Die totale Mobilmachung" and *Der Arbeiter*, left a powerful and lasting impression upon him. Jünger was deeply influenced by Nietzsche, and Heidegger saw Jünger's thinking as a continuation of Nietzsche's will to power. But he could not fathom, let alone acknowledge, Jünger's military heroism and "mystical" worship of technology. On the contrary, Jünger's supposed Nietzscheanism partly explains Heidegger's interpretation of Nietzsche's will to power as *Machenschaft*.

The concept of *Machenschaft* plays a key role in Heidegger's thinking about technology. The meaning of the word is close to "making," i.e. that which can "be

[26] Heidegger: *Nietzsche*, vol. III, p. 175.

made" or "produced." *Machenschaft* is technological power's "making" in a world of fully disposable, usable, and compliant people—a place where everything can be made or produced through manipulation and exploitation of nature and human beings.

One of the most important philosophical works of the postwar era, Heidegger's *Nietzsche* was published in 1961. This weighty volume on Nietzsche's philosophy is comprised of essays and lectures that Heidegger gave during the period 1935–1946, that is to say, shortly after his so-called "turn."

In the third volume of *Nietzsche*, Heidegger connects the concept of *Machenschaft* to *Being*.[27] Yet this is not about the originary truth of Being but about Being as total mobilization, of which the only "meaning" is utter senselessness. All that remains is the empty superficiality of a "life" in which power is desired for its own sake.

For Heidegger, the inauguration and maintenance of this utter senselessness represents the end of metaphysics. Metaphysics—as the history of Being—has thereby run its course from Plato to Nietzsche. To the extent that the Platonic *Idea* can be regarded as a "value," Nietzsche's thinking involves an inversion of Platonism, becoming a *reevaluation of all values*, what Nietzsche calls nihilism.

So as to offer a counterpoint to this critique of technology, Heidegger in several of his late writings has foregrounded other kinds of (positive) experience: we can ask questions of technology; we can experience joy in art and poetry; we can meditate freely; and we can adopt an attitude of relaxed calm, *Gelassenheit*. In the final instance, the core of Heidegger's philosophy of technology nonetheless remains the questioning of technology itself.

2.2.4 Authoritarian Technology

There are philosophers who consider modern technology as a curse rather than a blessing. Although he was not a phenomenologist, the American author Lewis Mumford shared phenomenologists' bleak view of modern technology. Mumford wrote prolifically on everything from architecture and metropolises to technology in general. He was something of a philosophical eccentric and never completed a university degree. His method was to work interdisciplinarily by combining history of technology, philosophy, literary and art criticism, architecture, anthropology, and sociology.

Mumford had a strongly humanistic bent and his sustained criticisms of modern technology followed in an American tradition of worldly romanticism that stretches from Ralph Waldo Emerson to John Dewey. This tradition is worldly in its concern for the environment, the peacefulness of small-town life, and the conservation of nature. It is romantic in its assertion that the basis for human activity is not material nature but personal will and humanity's quest for self-realization through creative activity.[28]

---

[27] Ibid.

[28] Mitcham, *Thinking Through Technology*, pp. 40–41.

Mumford's first book, *Technics and Civilization* (1934), is a wide-ranging account of the history of machine technology that also describes contemporary reactions, psychological and cultural, to that technology:

> We have observed the limitations the Western European imposed upon himself in order to create the machine and project it as a body outside his personal will: we have noted the limitations the machine has imposed upon men through the historic accidents that accompanied its development. We have seen the machine arise out of the denial of the organic and the living, and we have in turn marked the reaction of the organic and the living upon the machine.[29]

Thirty-five years later, Mumford published *The Myth of the Machine* (1967), which repeats his earlier position even while insisting more sharply that human beings should be regarded as *Homo sapiens*, not *Homo faber*. Mumford denies that humanity is primarily defined by an ability to make tools. The foundation of humanity is not making but thinking, not tools but personal will. Like many other twentieth-century thinkers, Mumford argues that language, or, more generally symbolic representation, rather than tools, which most obviously separates humans from other animals.[30] Other originary human qualities include the ability to interpret things and to create art.

Contrary to the common notion that abstraction and rationality are the basis for human progress, Mumford argues that dream states, with their frequently capricious associations of images and ideas, are the source of this originary creativity. In the youth of humanity, the human body itself was a crucial medium for this creativity. The body's role in dances and ceremonies, and as the object of body art, tattooing, and ornamentation cannot be overstated, Mumford argues. He even argues that the human body was the most important "tool" of early humans. He highlights the fact that women's production of things like baskets and clay pots were crucial for the social coherence of early human communities.

On the basis of this view of the essence of humanity, Mumford distinguishes between two forms of technology: a decentralized, small-scale version—what he calls "democratic technology"—and another that he regards as more centralized, top-down, and oppressive—"authoritarian technology." The latter, according to Mumford, is the dominant mode today. His vision is that authoritarian technology will be replaced by a more democratic technology—liberatory, egalitarian, life-affirming—that in many respects resembles the ecological thinking of our own era.

Authoritarian technology, according to Mumford, was not introduced as part of the Industrial Revolution; rather, it goes back more than five thousand years to the earliest civilizations. The first "machines"—which Mumford calls "megamachines"—were not made of wood or metal but consisted of organized labour in the

[29] Mumford, *Technics and Civilization,* p. 433.

[30] Mumford's conclusion is supported by recent zoological research showing that not only humans but also animals create tools. Chimpanzees and even insects and crabs seem to use twigs, pebbles, sponges, and other tools. However, humans are distinguished from other animals by their creativity and capacity to "devise tools to make tools to make more tools" (Dusek and Scharff, *Philosophy of Technology: The Technological Condition*, p. 342).

form of armies of workers gathered together to create colossal buildings and irrigation schemes. Examples of gigantic construction projects undertaken by the first civilizations include the Egyptian pyramids, the Great Wall of China, and the Sumerian canals of the Tigris-Euphrates delta.

These mega-machines often brought considerable material prosperity but at the price of severe limitations upon the individual's personal goals and desires. A large army can conquer territory to extend its ruler's power but only by forcing its soldiers to submit to a discipline that either puts an end to family life, play, poetry, music, and art or makes them subordinate in every aspect to military objectives. The result is "the myth of the machine"—the idea that megatechnology is both unstoppable and, ultimately, beneficial. It is a myth because it can be stopped and is, ultimately, not beneficial. Mumford's entire line of argument seeks to de-mythologize and limit megatechnology, and thereby set in motion a radical change in attitude that will lead to the transformation of authoritarian technology.

Mumford additionally argues that twentieth-century totalitarian states, but also modern democracies with their welfare state bureaucracies, massive military-industrial complexes, and secretive cadres of scientists and security experts, more closely resemble the earliest civilizations than the more decentralized market economies of the nineteenth century.

However, this is not to suggest that Mumford argues for the rejection of every—or even any—kind of technology. On the contrary, he tries to prove that we need to distinguish between two kinds of technology, of which one is in harmony with human nature and the other not.

Technology should be valued when it facilitates and enhances what Mumford calls "the personal" in our lives, not when it limits or constrains people's lives by putting power first. Or, in Mumford's own words: "to save technics itself we shall have to place limits on its heretofore unqualified expansion."[31]

2.2.5 *Autonomous Technology*

At the same time as Heidegger was working on the question of technology, Jacques Ellul was developing a systematic analysis of "technique"—*la technique*—as the most important social phenomenon of modernity. For Ellul, the dominant force in social development is no longer capital, as it was in Marx's day, but, rather, "technique," which Ellul define as "the *totality of methods rationally arrived at and [aiming at] absolute efficiency* (for a given stage of development) in *every* field of human activity."[32]

Ellul's concept of *la technique* refers not to a particular application of technology, nor to a particular machine, technology, or way of achieving a goal, but the

---

[31] Mumford, "Technics and the Future of Western Civilization," p. 39.

[32] Ellul, *The Technological Society*, p. xxv.

application of technology within *every* sphere of life and society. For Ellul, *la technique* is more precisely a perspective, mentality, or a kind of intentionality—a new mode of being-in-the-world.

In *La Technique* (1954), Ellul distinguishes between "technical operations" and "the technical phenomenon." Technical operations are by their nature plural, traditional, and context-dependent, while the technical phenomenon —*la technique*—is indivisible and underpins the modern way of producing and using artifacts, a way that tends to dominate and absorb all human activity. With this all-encompassing pursuit of effectivity, "technique has taken over the totality of human activities, not only those of productive activity."[33] Ellul describes the technical phenomenon in terms of seven defining features in all modern technology: rationality, artificiality, automatism, inbuilt growth, indivisibility, universality, and autonomy.

Modern technology replaces the natural environment with an increasing degree of rational human activity. It constantly expands in accordance with its own inner dynamic. To quote a somewhat overused saying: "The solution to the problems of technology is not less but more technology." Yet it happens only rarely that one technology is phased out when another is introduced. Modern technology is not merely everywhere expanding, it is also becoming increasingly uniform across the board; it lives its own life, it becomes *autonomous*. When Ellul states that modern technique has become autonomous, he means that technological development is overwhelmingly predetermined and fixed. Technique now rules people, not the other way round. Technique lives its own life independently of people's efforts to steer it towards concrete objectives.

The goals of technology are often described in terms of progress, happiness, prosperity, and so forth, concepts that are usually devoid of meaning. The goals are vacuous and meaningless even as the means become increasingly complex and difficult to understand. Engineers, businesspeople, and politicians are all unable to understand the consequences of the technology that they are developing or advocating. While engineers in general are ignorant or naïve about the social consequences of the technology they are developing and marketing, businesspeople and politicians are even more ignorant about how technology in itself works. The public understands the technological and social consequences of technology least of all. It is also difficult to question abstract goals such as prosperity and freedom, for which reason the means tend to become more important than the goals. And the goals are ascribed a diffuse significance that is adapted to the means available. Society adapts itself to technology, rather than technology adapting itself to society.

For Ellul, autonomous technology, to the degree that it exists, is not predetermined, as in Heidegger. Rather, it is what the subtitle of his book calls *L'enjeu du siècle*—the bet of the century. The rise of modern technology is not some blind conquest of nature, it is our deliberate transformation of a natural world into a technological world. Much is at stake in all this. Either we succeed in creating a better world (if such a thing is even possible) or we fail. Everything depends on our ability

[33] Ellul, *La Technique, ou L'Enjeu de siècle*, p. 2.

to think, manage, and act with good intentions. And in this colossal "lottery" there are no guarantees.

Ellul's thinking about technology may be closer to Heidegger than Ortega, but where Heidegger advocates *Gelassenheit*, Ellul argues for an ethics, an ethics of nonpower (*une ethique de non-puissance*):

> An ethics of nonpower—the root of the affair—is obviously that human beings agree not to do everything they are able to do. Nevertheless, there is no more … divine law to oppose technology from the outside. It is thus necessary to examine technology from the inside and to recognize the impossibility of living with it, indeed of just living, if one does not practice an ethics of nonpower. This is the fundamental option. … [We] must search systematically and willingly for nonpower, which of course does not mean accepting impotence …, fate, passivity, etc.[34]

Clearly, Ellul's ethics of nonpower is not without its problems. It will try not merely to set boundaries but also strive towards freedom from technology as an ideology, posing a challenge to powerful interests that risks giving rise to acute tensions and potential conflicts in the world of technology. If this ethics is applied fully, it will radically reduce the power of information technologies and the dominance of airplanes and cars as means of transportation, and it will reduce overconsumption and environmental destruction, which could in turn lead to a new way of talking and listening, building and living, and, indeed, perhaps even a new way of thinking that, in turn, might promote not only the freedom to question things but also a particular counter-technological alternative. At the same time, it may be asked whether a technology unconstrained by power is possible without the human driving forces that presumably lie behind most of the technological systems and structures of modern society, namely the will to power, prestige, and prosperity that serve as incentives for countless technicians, scientists, and others in positions of power.[35]

---

[34] Ellul, "Recherche pour une Ethique dans une société technicienne," p. 16 (cited in Mitcham, *Thinking Through Technology*, p. 61).

[35] Perhaps the most thorough discussion of the ethical problems associated with technological power is Hans Jonas's *The Imperative of Responsibility* (1984). For Jonas, the problem with new technology is that its long-term consequences for the global environment are not always possible to predict. His thesis is that "the new kinds and dimensions of action require a commensurate ethics of foresight and responsibility which is as novel as the eventualities which it must meet" (p. 18). This new imperative, the imperative of responsibility, demands in turn a new kind of humility in the face of humanity's new-found power. Jonas's position is close to Ellul's argument for an ethics free from technological power, albeit an ethics whose rationale is theological rather than secular. Despite this nuance, both have set the terms for an ethical debate about technology that has affected, and continues to affect, both philosophers and the public at large.

2.2.6 Technological Power

Technocracy, or rule by experts, is a concept with a long history and in modern society it takes many forms. In antiquity, Plato strongly emphasized the significance of knowledge for ruling a state. Bacon asserted the power of knowledge over nature, and he described a utopia in which natural philosophers supplied rulers with knowledge and where their knowledge led to more power being vested in the state. This development was deeply criticized not only by Mumford and Ellul but also Michel Foucault. Throughout his life, Foucault voiced concern about the growth of scientific knowledge and its creeping dominance; his concern was also the driving force behind his investigations into the covert and open relations of power inherent to the "discourse of knowledge" and the political consequences of those relations for human life. His investigation of the origins of institutions, their legitimacy, and their social importance form the very core of Foucault's analyses of power and its technologies.

According to Foucault, "technology" cannot be limited to "the production and use of artifacts." Instead, the concept needs to be broadened to include what might be called technologies of behaviour. He lists four types of technologies: technologies of production, technologies of sign systems, technologies of power, and technologies of the self. The first two of these are usually used within the natural sciences and linguistics. The last two, *technologies of power* and *technologies of the self*, are the particular focus of Foucault's interest.[36]

In essays such as *Governmentality* (1978) and *Political Technologies of Individuals* (1982), Foucault investigates the technologies of power "which determine the conduct of individuals and submit them to certain ends or domination."[37] For example, he considers the rule of a state as an "art," a technology that follows certain rules. Only in exceptional instances does the state—"the coldest of all cold monsters"—need to care about a particular individual; the state and its long arm—the police—are a political technology that is founded on rationality in the same way as science is founded on rationality.[38] But Foucault was not especially interested in analysing the traditional opposition between ruler and rules, such as an absolute monarch and his subjects. Instead, he analyses relations of power between particular human institutions such as prisons, schools, asylums and hospitals, factories, and cities. Prisons, for example, were established in order to replace the earlier system of executions, which was regarded as inhumane. The result was the creation of "whole army of technicians took over from the executioner."[39]

A whole army of technicians soon began organizing an efficient penitentiary system based on Jeremy Bentham's vision of "the all-seeing eye"—the panopticon. Bentham's model institution was conceived as the perfect prison, allowing

[36] Foucault, *Technologies of the Self*, p. 18.

[37] Ibid., p. 18.

[38] Foucault, "Political Technology of Individuals," p. 161.

[39] Foucault, *Discipline and Punish*, p. 11.

continual surveillance of completely separated individuals. The result was an entirely controlled environment in which those surveilling were able to observe everything that took place. In such a panopticon, power lies in the capacity to actively promote an internalization of behaviour. Because prisoners feel that they are under a continual surveillance that they have no possibility of avoiding, the only strategy is to adapt. But even those surveilling are victims of this system: they are forced to observe the prisoners, at least now and then.

In this entire panopticon "machine," no single person has total control; all are subordinate to the workings of the machine. For Foucault, the panopticon is prototypical of the ideal system of power. The panoptical system becomes a "penitentiary technique" equal in importance to the techniques developed by the Inquisition.[40]

The panopticon became a model for the new prisons of the nineteenth century but also for the new relations of power in modern society, relations that collectively form the system of power that Foucault calls *discipline*. Disciplinary practices were introduced in order to create order in the system and to increase the usefulness and malleability of its component elements. Discipline increasingly became "a technique for making useful individuals."[41] In modern society, each of us is subjected to disciplinary power in the form of customary practices, behaviours, and ways of being in which we find ourselves for various reasons. Power expresses itself in a practice of self-regulation that is based on a number of self-imposed rules, which we try to follow so that society can function. Foucault refers to this self-imposed mode of subjugation at the individual level as a "technology of the self"

In his essay *Technologies of the Self* (1982), Foucault discusses a series of technologies of the self that have been advocated and exercised from antiquity to the present day. He shows that technologies of the self have their historical roots in two separate movements: Graeco-Roman philosophy and Christian spirituality. Graeco-Roman philosophers referred to "***epimelesthai sautou***, 'to take care of yourself," "the concern with self," "to be concerned to take good care of yourself'."[42] This care of the self—a form of technology of the self—was connected to the Delphic motto "Know thyself," over which it also took precedence. By contrast, the Christian moral tradition emphasized self-denial as a condition of salvation.

Yet Christianity is not unique in promoting self-denial, which also features in other religions and philosophical traditions. The various technologies that Foucault considers include the *ascesis* (asceticism) of the Stoics, which means not self-denial in the Christian sense but rather "the progressive consideration of self, or mastery over oneself, obtained not through the renunciation of reality but through the acquisition and assimilation of truth."[43] As heirs to the Christian moral tradition, we are hardly predisposed to apply *ascesis*, being more inclined to follow the path of self-denial, to suppress our own desires, and to obey power.

---

[40] Ibid., pp. 226, 253, 255.

[41] Ibid, p. 211.

[42] Foucault, *Technologies of the Self*, p. 19.

[43] Ibid., p. 35.

Generally, Foucault used the word "power" in preference to "technology" as a designation for his philosophical approach as he was primarily interested in the relationship between power and knowledge, a concept that is directly related to the word "technology." The slight difference in terminology is relatively unimportant in this context. In fact, Foucault's treatment of the concept of power and his frequent use of the words "technique" and "technology" have a lot in common with the broader definitions of "technology" that we find in Mumford, Ellul, and Heidegger. Foucault clearly shares the same fundamental position as Mumford, whose "megamachine" is a technological apparatus comprised of people, and as Ellul, who refuses to conflate "*la technique*" with a particular form of hardware, preferring instead to view it as a rationalization of every dimension of life for the purposes of maximizing efficiency. And we can agree with Andrew Feenberg that there are "significant similarities between Heidegger's critique of technology and Foucault's writings on power, especially in the period of *Discipline and Punish*."[44] Although there are clear parallels between Foucault and these other thinkers, it is nonetheless in the field of empirical studies of science and technology that Foucault's philosophy of technology has had the greatest impact.

In the next section, I address a line of thinking that seeks to overcome what it sees as the obvious limitations in the philosophy described above. Instead of viewing technology as a kind of framework beyond the individual technological artifacts, this line of thinking proceeds from the empirical reality of technology, that is to say, the technological artifacts and objects themselves. I begin by summarizing the empirical scholarship on technology and science, after which I discuss an attempt to formulate an alternative to phenomenology: postphenomenology.

2.3 Postphenomenology and Empirical Philosophy of Technology

2.3.1 Empirical Science and Technology Studies

The social and cultural basis for technological-scientific research have become a focus of study in recent years. Researchers in the field known in the English-speaking world as Science and Technology Studies (STS) undertake empirical studies of different aspects of the interaction between technology, science, and society. STS is a broad, multi-disciplinary, and fast-growing research area that includes disciplines such as sociology of science, history of science and technology, and theory of science. STS researchers use existing theories and methods in the social sciences and the humanities in order to find out how engineers, technicians, and scientists work in practice. They study these professionals as *people*, observe them in their

[44] Feenberg, *Questioning Technology*, pp. 6–7.

concrete work settings, interpret their documents, and try to explain their actions in terms of current theories about human behaviour.

Empirical research has been and remains openly critical of phenomenology and transcendental philosophy of technology. The fundamental weaknesses identified by empirical research is the generalizing and misleading account of technological determinism that it sees in the work of Heidegger, Ellul, and others. For these latter, technology developed autonomously and society adapted itself thereafter. Yet empirical research on concrete technologies and their interactions with society undermined this deterministic view. Instead, technology was revealed as being not an autonomous craft but the result of human activity. Its development was primarily dependent on the choices that people made in the special situations in which they happened to find themselves. The development of technology should therefore be regarded as *socially constructed*, or as simply *constructed*.

A key question in this context is whether there has been an explicit philosophical reflection upon the STS's own foundations. Going through the huge literature on STS, it is hard not to be struck by the striking absence, indeed, outright rejection, of philosophical thinking. An interesting exception is Steve Fuller's *The Philosophy of Science and Technology Studies* (2006). Fuller argues that as the field has become more established, it has become steadily detached from its philosophical roots, and that it is vital that this contact be reestablished.

According to Fuller, one reason for STS's philosophical rootlessness is that technology and science are now inextricably intertwined, forming an undifferentiated technological-scientific culture of *technoscience* in combination with the society around them. Making this nebulous concept into a focus for philosophical reflection has proven difficult.[45]

Another reason is that professional philosophers are regarded as an obstacle to constructive work because they continually ask awkward questions without any thought as to the practical consequences. Fuller is here primarily criticizing "analytic philosophers," whereas "Continental philosophers" are seen as more amenable to STS's theories and methods, for the simple reason that they consider STS as applications or extensions of ideas already implicit in the thinking of Heidegger, Foucault, or Deleuze. However, this does not prevent STS (and Fuller) from criticizing what they see as the condescending attitude of Continental philosophers, who generally treat STS researchers as a species of "speculative sociologists" whose limited feeling for context prevents them from reaching beyond their own "intertextuality."[46]

For Fuller, STS's historical resistance to philosophy can primarily be explained by the fact that most of STS's founders were scientists. They therefore had an obvious (and perhaps unconscious) inclination towards positivism, which, according to Fuller, is the philosophical origin of STS and, in particular, the sociology of knowledge.

---

[45] Fuller, *The Philosophy of Science and Technology Studies*, p. 4.

[46] Ibid., p. 7.

What Fuller tries to show is that STS should be regarded as a postmodern version of positivism. In discussions of positivism, a distinction is usually made between the "positivistic" movement founded by Auguste Comte in the 1830s and the more strictly philosophical movement that is known as *logical positivism* and that is closely bound up with the Vienna Circle of the 1930s. What both these movements have in common, however, is that they warned of an uncontrolled rationalism with potentially catastrophic practical consequences. Both therefore argued that the foundation of positivism must be solidly *empirical*.[47] Fuller thereby differentiates in practice between philosophy and STS, ascribing philosophy a normative role and STS a descriptive role.

2.3.2 STS and Actor-Network Theory

The (social) constructivism of STS and its techno-scientific approach essentially means abandoning philosophy as a *normative* project. Philosophy thereby goes from being an "imperious legislator" to what John Locke called an *under-labourer*, a word that he coined to describe his own intellectual subordination to Isaac Newton. *Under-labouring*, as represented by Leibniz, for example, can be regarded as the kind of "uncritical" philosophizing that Kant called "dogmatic." This means that what philosophers in general treat as *problems*, STS—in the spirit of Wittgenstein's later writings—treats as *facts*. For STS, the existence of facts is unproblematic under normal circumstances, once they have been constructed. There is practically no room here in which to ask philosophical questions about the fundamental basis of technological-scientific research; the primary task of the STS is simply to make observations and *describe* them without "epistemological prejudices."[48]

STS accordingly starts from different *theories* rather than from philosophy. "Theory" in this context is a number of possible frameworks that set out how STS research should be conducted. The dominant theory, Actor-Network Theory (ANT), is closely identified with the philosopher Bruno Latour, the engineer Michel Callon, and the sociologist John Law. The theory was developed in France in the early 1980s in an attempt to understand large technological-scientific systems and their place within a wider social context. For Actor-Network Theory, the participants include not only human actors but also non-human actors such as laboratory animals and dead objects. The former should not be given priority over the latter. Latour calls the participants in networks *actants*, which may be *humans* or *non-humans*. Within the network, behaviours cannot be explained in terms of human "subjects" or non-human "objects" but are instead the result of heterogeneous relations between humans and non-humans.

---

[47] Ibid., p. 79.

[48] Ibid., pp. 45–6.

Fuller argues that ANT is equally opposed to classical liberalism in the style of John Stuart Mill and to revolutionary Marxism. He views each of these traditions as advocating, on its own terms, economic policies that are based on overarching hierarchical structures and that seek to stifle, or fully prevent, the free movement of goods and services, people and money. He also regards them as seeking to impose restrictions upon people's freedom to draw up agreements for their own mutual benefit. Fuller sees ANT, in sharp contrast to these two intellectual traditions, as supporting scientific policy that accords with the neoliberalism of the last few decades.[49]

Several objections can be raised to Actor-Network Theory. For example, it can be criticized as amoral. It does not define how people *should* act ethically and politically. Advocates for ANT would likely answer that while adopting an ethical and political position is entirely possible, it must always be preceded by a *description* of the actor network.

The theory can also be criticized for ascribing the same importance to every actor in the network. In my view, adherents of ANT tend to ignore the presence of preexisting (power) structures when describing how these structures emerge from the interactions of actors in the network and through the actors' capacity to promote their own interests.[50] Finally, it can be argued that actor-networks risk degenerating into unending chains of association, some ultimate vast network into which we are all incorporated.

2.3.3 Constructivism and Amodernism

One of the most influential representatives for STS is the French philosopher and anthropologist Bruno Latour. In *Laboratory Life* (1979), Latour and his co-author Steve Woolgar introduce a new "anthropological" approach to social science, showing by means of case studies how daily activities in the laboratory lead to the surprising conclusion that *scientific facts are constructions*. The book, which partly draws on ideas taken from philosophers such as Gilles Deleuze, Michel Foucault, and Donna Haraway, also served as an inspiration for the development of Actor-Network Theory. It is also worth noting that Latour used Derrida and deconstruction, though without naming either, in order to undermine a number of hierarchical structures, such as those connecting technology to science and nature to culture.[51]

In *We Have Never Been Modern* (1993), perhaps his most philosophical work, Latour asks the question of what it means to be modern. What role is played by

[49] Ibid., p. 67.

[50] See, for example, Callon, "Some Elements of a Sociology of Translation," pp. 196–229.

[51] Deconstruction can be defined in many ways but common to all is that they start from an understanding that the philosophical tradition is characterized by thinking in hierarchically structured binary opposites, such as speech/writing, male/female, central/marginal, in which the first term dominates over the second. Deconstruction questions this dominance by inverting the order and erasing the polarity of the terms. Its aim is to destabilize established philosophical positions and hierarchies in the hope that new perspectives and insights will emerge.

modern science? According to Latour, our predecessors would have been baffled by the sharp distinctions that our own era makes between nature and culture, thought and things, subject and object. However, in parallel with modern science's quest for differences, we are also now seeing a contrary tendency: the construction of hybrid solutions through the integration of technology and politics, science and society. Recent examples of these integrative efforts include debates about the ozone hole, acid rain, and global warming. The more integration progresses, the more hopeless seems the effort to preserve humans and non-humans as two separate domains. Instead, Latour argues, we should redefine modernity and recognize that the two domains are intimately intertwined.

Pandora's Hope (1999) is an extension and an updating of *We Have Never Been Modern.* The book is a collection of essays in which Latour returns to the relation between humans and non-humans, people and their world. Latour's primary objective in the book is to criticize Cartesian dualism: "How could we avoid noticing that we have not moved an inch since Descartes?"[52] According to Latour, Cartesian dualism is a hallmark of both modernists and postmodernists, each of whose approaches relies on a faulty assumption, namely that humanity is separated from its world by an absolute ontological chasm. Both modernist realism and postmodernist relativism presuppose that there is a chasm separating the thinking subject from the world around it. Once this chasm has been accepted, the crucial question then becomes how to bridge it. For the realist, the bridge is science; for the relativist, it is the play of language. For Latour, however, there simply is no chasm!

What Latour tries to show here is that it is futile to try to reduce this unbridgeable gap by seeking a correspondence between consciousness and things. Drawing on case studies on the Amazonian rainforest *inter alia*, he proposes an entirely different solution that calls for a number of tiny steps, each of which is a "translator" that translates between matter and form. A relatively small gap separates every step from the one that comes after it. The translators are connected by a chain stretching all the way from things to consciousness, or, to use Latour's terminology, from non-humans to humans. Unlike the Cartesian model, Latour's model is *unlimited* and *reversible*. Facts travel in both directions, *circulating* back and forward like electricity in a cable that is open at both ends. In such a model, all action is distributed along chains of human and non-human actants, which collectively form an actor network.

The main thesis of *Pandora's Hope* is unambiguously philosophical: Latour wishes to settle accounts with what he regards as the artificiality of modernist and postmodernist epistemology:

> There is no world outside, not because there is no world at all, but because there is no mind inside, no prisoner of language with nothing to rely on but the narrow pathway of logic. … Epistemology was never intended to protect [the scientists and the engineers], it was always a war machine."[53]

[52] Latour, *Pandora's Hope*, p. 8.

[53] Ibid., p. 296.

Epistemology is science and technology's war machine because it shields the scientific elite from the rest of society: it is a Foucauldian power politics, a politics that sets itself above politics. Latour's alternative to this "warlike" worldview, which seems increasingly incapable of coping with its own advances, is to allow the non-human and the human to unite on equal terms in a collective whole.

2.3.4 Postphenomenology

Philosophy of technology today spans a great many areas. A good way to see what current debates are focusing on is to read three conference catalogues for the Society for Technology and Philosophy (SPT). The theme of its 2009 conference was Converging Technologies, Changing Societies, a reference to the increasingly rapid convergence of information technology, biotechnology, nanotechnology, and cognitive technology and the societal consequences of that convergence. The theme of its next conference, in 2011, was Technology and Security, with "security" being used in the widest sense of the term, including ecological, social, and economic sustainability as well as military, cyber-, and national security. Its 2013 conference, which was held in Lisbon and brought together more than 300 participants from thirty countries, had the theme of Technology in the Information Age. The aim was to "stimulate debate surrounding the cultural, social, economic, political and ethical consequences of the progress within IT and media technology."

These conference themes addressed general questions about technology's relation to different trans-technological areas such as ethics, politics, democracy, religion, globalization, the public sphere, surveillance, critical thinking, and phenomenology. As we will shortly see, these are precisely the questions that Derrida takes up in every one of his writings. In other words, Derrida seems to be at the very centre of the current debate within philosophy of technology.

This debate is today dominated by so-called "postphenomenology." It was formulated by Don Ihde and Peter-Paul Verbeek as an alternative to Heidegger and Ellul's portrayal of technology as a form of alienation: it alienates human beings from themselves by preventing them from living authentically. According to postphenomenology, modern philosophy of technology must do justice to the empirical aspect of technology without abandoning the hermeneutic and existential questions asked by phenomenologically oriented philosophy of technology. It therefore tries to shift the focus of attention from technology's conditions of possibility to technology itself—to the technological apparatuses and artifacts that everywhere surround us in our daily lives—in order thereby to try to understand their role in our society and culture.

The word "postphenomenology" was coined by Don Ihde, a leading North American philosopher of technology and the author of the first English-language studies of philosophy of technology, *Technics and Praxis: A Philosophy of Technology* (1979). In his book, Ihde distinguishes between an idealist and a materialist approach to technology. The former regards technology as applied science, while the latter sees science as theoretical technology. In aligning himself with the latter position, Ihde is able to sketch the outlines of a "postphenomenological" analysis of the humanity–technology–world relation.

Ihde's aim is to enhance this relation by adding a hermeneutic dimension, which he believes can offer a very different perspective on technology than the hermeneutic perspective developed by Heidegger. Although both ask the same question—the question of the meaning of technology—they therefore produce entirely different results. Where Heidegger fundamentally regards technology as the way in which the whole meaning of the world is unveiled, Ihde focuses on the hermeneutic role of the various technologies. What separates Heidegger and Ihde is the obvious difference in their respective views of technology. Heidegger regards technology as a way of interpreting the world—indeed, ultimately, as *the* way of interpreting to which Western culture is committed. By contrast, Ihde's understanding of technology starts not with such a world-interpretation but with our relations to concrete technological artifacts and the ways in which we can interpret them. When the question of meaning is posed from this perspective, an entirely different image of technology emerges. It is concrete technologies, Ihde argues, that define the relation between humanity and its world and that make possible the various ways in which humanity can experience it.

Ihde's postphenomenology was further developed and radicalized by Peter-Paul Verbeek. Within a very short space of time, Verbeek has established himself as one of the most distinguished philosophers of technology. Part of a younger generation, he can be thought of as Ihde's postphenomenological heir. Verbeek's most important book is *What Things Do* (2005), in which he distances himself from Husserl, Heidegger, Jaspers, and others, who, he argues, articulate a partly unfounded fear that technology is cutting humanity off from itself and the world around it. In contrast to what Verbeek sees as this overly bleak and abstract view, he builds on the work of Ihde, Latour, and others by presenting an empirically richer account of how material things (such as TV sets, cars, and mobile phones) shape our lives and experiences.

A radical version of postphenomenology is so-called "transhumanism." While it has a long prehistory, in its current form it can be defined as "an intellectual and cultural movement that aims to transform the human condition by developing and creating widely available sophisticated technologies to greatly enhance human intellectual, physical, and psychological capacities."[54] Its goal is thus to create a new, transhuman life-form that eclipses *Homo sapiens* in virtually every area. According to transhumanists, such an improvement of the human race is not only inevitable but highly desirable. Probably the leading transhumanist today is Nick Bostrom (b.1973), a visionary and unconventional philosopher whose stated goal is to think rationally about the big questions facing humanity. In order to promote the fast-growing transhumanist movement, he joined with David Pearce in 1998 to found the World Transhumanist Association, an international umbrella organization

[54] World Transhumanist Association: Transhumanist Declaration.

for transhumanism that currently has around 5000 members in more than a hundred countries.

Even though many philosophers of technology, including Verbeek himself, are arguing for an enhancement of humanity by means of technology, a great deal of criticism has predictably been levelled at the transhuman credo. For example, is it meaningful to speak of an "enhancement" of humans' *emotional life*? The uncritical and unchecked use of technology to improve human emotions can reasonably be said to show a lack of insight into the complex and subtle nature of emotional life. For instance, it is far from certain that strengthening a positive feeling such as "happiness" leads to a good or more "authentic" emotional life. An even more serious objection is that large-scale changes to our emotional life—as Martha Nussbaum and others have noted—entail drastic changes to our cultural norms and values. For example, should people experience profound sadness in connection with dramatic events when effective anti-depressant medications are available? The critique of transhumanism is well summarized by Andrew Feenberg, who simply points out that humans are infinitely complex beings whom we have barely even begun to understand.

I want to turn finally to Bernard Stiegler, one of the most original and thought-provoking contemporary philosophers of technology, who was also Derrida's student and perhaps his most important debating partner. Stiegler's thinking, which I regard as having much in common with postphenomenology, recurs in various ways in this study, for which reason I will here limit myself to a few of his main ideas. In his most famous work, *La technique et le temps* (published in three volumes in 1994, 1996, and 2001), he advances his thesis that humanity should be thought of as *Homo faber* and technology as the gradual *externalization of humanity*. Technology is instrumental and encompasses everything from primitive tools to various writing systems and modern telecommunications. However, it is not simply instrumental but also determines what it is to be human: the "human" is always already grounded in technology. Technology should be seen, in fact, as a prosthesis: human beings should not primarily be viewed as *Homo sapiens*. Fundamentally, we are not rational creatures but exist in a prosthetic relation to matter, what Stiegler calls "organized inorganic matter." According to Stiegler, this prosthetic externalization of human beings involves a radical break with genetic evolution and the start of human development, which is a development towards increasingly advances technological aids. For Stiegler, technology is a "pursuit of life by means other than life."[55]

But Stiegler is not simply searching for a new way to describe the relation between human beings and technology. Rather, he wants to show that technology poses a challenge to philosophy itself. Indeed, Stiegler claims that it is highly dubious to speak of "philosophy of technology" because "since its origin, technology has endured this technological condition, *but as repression and denial*, and that is the entire difficulty of my undertaking—to show that *philosophy begins with the*

[55] Stiegler, *Technics and Time 1*, p. 17.

repression of its proper question" [original emphasis].[56] As Stiegler here suggests, technology cannot be understood solely as a history of technological development; rather, it must be thought of primarily as a deconstruction of philosophy. In this spirit, Stiegler offers a series of "deconstructive" readings of Rousseau, Kant, Husserl, Heidegger, and others, in which he shows how the "technological presupposition" is suppressed in their writing.

As a philosopher of technology, Stiegler is the true heir of his mentor Jacques Derrida, even if his analyses tend towards a positivistic understanding of technology that Derrida himself does not share.

2.4 Summary

Perhaps the most surprising aspect of Derrida's ideas about technology is how little contact they have had with the larger philosophy of technology tradition. With the exception of Husserl, Heidegger, Marx, and Stiegler, his more than one hundred works make almost no reference to other philosophers of technology. As far as I have been able to establish, he makes no reference to Kapp or Dessauer or to philosophers such as Ortega, Mumford, or Ellul. Social philosophers, sociologists, postphenomenologists, and transhumanists are largely absent from his texts. Foucault's name appears sporadically, and those of representatives of the Frankfurt School even more so.[57]

However, despite this apparent lack of contact with large parts of the philosophy of technology tradition, elements of Derrida's thinking exhibit similarities to particular ideas in that tradition. There is undoubtedly a materialistic tendency in his writing that has clear parallels with Kapp's grounded materialism and prosthetic thinking. There is also a kind of analogy between Derrida's definition of "technology" and Kapp's comparison of external technology and what is inside a human being, as when he represents telegraphy as an externalization of the human nervous system.

Derrida's philosophy of technology also shares certain features with the empiricism of Science and Technology Studies. However, as we have seen, rather than Derrida himself being influenced by STS, the latter has been influenced by his thinking, as when its leading representative, Bruno Latour, used deconstruction to undermine a number of hierarchically structured binary opposites such as technology/science, nature/culture, and human/non-human.

---

[56] Stiegler, *Philosopher par accident*, pp. 14–15.

[57] As far as I am aware, although Derrida had a protracted and highly publicized controversy at an abstract philosophical level regarding the aims and contents of Foucault's *Histoire de la folie* (1961), Derrida never commented on Foucault's work on technologies of power and technologies of the self. In fact, he has written surprisingly little about technological power and power relations in society. Rather, his main interest is in which Foucault calls production technologies and technologies of sign systems, areas that Foucault himself did not engage with.

In this context, it is also important to emphasize Derrida's strong interest in developments within the natural sciences. He closely followed the latest research in this area, and he actively sought inspiration in the rapid advances being made in biology, genetics, anthropology, and cognitive science as well as in the new sciences of information technology and cybernetics.

Derrida's philosophy of technology also has a great deal in common with Ihde's and Verbeek's postphenomenology, for example, in their resistance to one-sided understandings of essence and their view of technology as a *relation* between humans and their world. Despite these superficial similarities, however, they differ on important issues. First, Derrida claims that empirical technology is what he calls a *supplement* to human beings, a prosthesis or (in Kapp's terms) an organ projection. It is a necessary supplement, but it is not, as Verbeek argues, constitutive of the human.[58] Second, Derrida's thinking is antithetical to a postphenomenology that has strong connections to American pragmatism and its frequently one-sided critiques of technology's critics. Third, given the significance that postphenomenology ascribes to technology itself, the question arises whether, in practice, it is not simply a sophisticated system of empirical thinking, which insufficiently addresses technology's transcendental aspect.

But even if Derrida has much in common with the philosophers who emphasize the "material" and empirical aspects of technology, he remains deeply rooted in a philosophical tradition that seeks to understand modern technology from a hermeneutical-philosophical perspective. This does not mean that he is always in agreement with the views expressed by its representatives or, for that matter, by other philosophers; on the contrary, his deconstructive analyses of other people's thinking about technology and science are often skeptical and problematizing. For example, in his essay *Ousia et Grammē* and his *magnum opus* on friendship, *Politiques de l'amitié*, he deconstructs Aristotle's notion of *techne* and its relation to time, politics, and democracy. In *De la grammatologie*, he shows how the ostensible opposition between nature and culture in the writings of Jean-Jacques Rousseau and Claude Lévi-Strauss are dissolved by *writing* as both technological-empirical fact and general structure. In *La voix et le phénomène* (1967), he calls into question Husserl's and the entire philosophical tradition's so-called "metaphysics of presence," and in *Échographies de la télévision* and other works, he problematizes Heidegger's theories of Being and its detachment from technology itself. Even so, Derrida's own ideas about technology are precisely based upon his lifelong and productive dialogue with Husserl's and Heidegger's writings. In the next chapter, we will see how this line of thinking emerges, what Derrida understands by technology, and what is unique about his own philosophy of technology.

58 Verbeek, *What Things Do*, pp. 112–3.

References

Callon, M. (1986). Some elements of a sociology of translation: Domestication of the scallops and the fisherman of St Brieux Bay. In J. Law (Ed.), *Power, action, and belief* (pp. 196–229). Routledge and Kegan Paul.

Cassirer, E. (1930). Form und Technik. In L. Kestenberg (Ed.), *Kunst und Technik* (pp. 15–61). Wegweiser.

Cassirer, E. (2012). Form and technology (trans. W.M. Dunlavey & J.M. Krois). In A. S. Hoed & I. Folkvord (Eds.), *Ernst Cassirer on form and technology: Contemporary readings* (pp. 15–53). Palgrave Macmillan.

Dessauer, F. (1927). *Philosophie der Technik*. F. Cohen.

Dessauer, F. (1956). *Streit um die Technik*. J. Knecht.

Dusek, V., & Scharff, R. C. (Eds.). (2003). *Philosophy of technology: The technological condition: An anthology*. Blackwell Publishing Ltd..

Dusek, V. (2006). *Philosophy of technology: An introduction*. Blackwell Publishing Ltd..

Ellul, J. (1954). *La technique, ou L'Enjeu du siècle*. Colin.

Ellul, J. (1964). *The technological society*. (trans. J. Wilkinson). Knopf.

Ellul, J. (1983). Recherche pour une Ethique dans une société technicienne (pp. 7–20). In Hottois (Ed.), *Ethique et Technique*.

Feenberg, A. (1999). *Questioning technology*. Routledge.

Foucault, M. (1961). *Histoire de la folie à l'âge classique*. Plon.

Foucault, M. (1977). *Discipline and punish*. (trans. A. Sheridan). Pantheon.

Foucault, M. (1988). The political technology of individuals. In L. H. Martin, H. Gutman, & P. H. Hutton (Eds.), *Technologies of the self: A seminar with Michel Foucault*. University of Massachusetts Press.

Foucault, M. (1991). Governmentality (trans. R. Brandotti and revised by C. Gordon). In G. Burchell, C. Gordon, & P. Miller (Eds.), *The Foucault effect, studies in governmentality* (pp. 87–104). University of Chicago Press.

Fuller, S. (2006). *The philosophy of science and technology studies*. Routledge.

Heidegger, M. (1954). Die Frage nach der Technik. In *Vorträge und Aufsätze*. Neske.

Heidegger, M. (1962a). *Being and time*. (trans. J. Macquarrie & E. Robinson). Harper and Row.

Heidegger, M. (1962b). *Die Technik und die Kehre*. Neske.

Heidegger, M. (1971). The origin of the work of art. In *Poetry, language, thought*. (trans. A. Hofstadter). Harper & Row.

Heidegger, M. (1977). *The question concerning technology and other essays*. (trans. W. Lovitt). Harper and Row.

Heidegger, M. (1979–1987). *Nietzsche*, Vol. 1-4 (trans. D.F. Krell). Harper One.

Hoed, A. S., & Folkvord, I. (2012). *Ernst Cassirer on form and technology: Contemporary readings*. Palgrave Macmillan.

Husserl, E. (1970). *The crisis of European sciences and transcendental phenomenology*. (trans. D. Carr). Northwestern University Press.

Husserl, E. (1978). The origin of geometry (trans. D. Carr). In J. Derrida (Ed.), *Edmund Husserl's origin of geometry* (pp. 155–180). University of Nebraska Press.

Ihde, D. (1979). *Technics and praxis: A philosophy of technology*. D. Reidel.

Jonas, H. (1984). *The imperative of responsibility: In search of an ethics for the technological age*. University of Chicago Press.

Jűnger, E. (1931). *Die totale Mobilmachung*. Verlag für Zeitkritik.

Jűnger, E. (1932). *Der Arbeiter: Herrschaft und Gestalt*. Hanseatische Verlag.

Kapp, E. (1877). *Grundlinien einer Philosophie der Technik: Zur Entstehungsgeschichte der Cultur aus neuen Gesichtspunkten*. [Fundamentals of a philosophy of technology: The genesis of culture from a new perspective]. Westermann.

Krois, J. M. (2012). The age of complete mechanization. In A. S. Hoed & I. Folkvord (Eds.), *Ernst Cassirer on form and technology: Contemporary readings* (pp. 54–61). Palgrave Macmillan.

Latour, B. (1993). *We have never been modern.* (trans. C. Porter). Harvard University Press.
Latour, B. (1999). *Pandora's hope: Essays on the reality of science studies*. Harvard University Press.
Latour, B., & Woolgar, S. (1979). *Laboratory life: The construction of scientific facts*. Sage.
Mitcham, C. (1994). *Thinking through technology: The path between engineering and philosophy*. The University of Chicago Press.
Mumford, L. (1934). *Technics and civilization*. Harcourt Brace. Reprinted 1963.
Mumford, L. (1954). Technics and the future of Western civilization. In *In the name of sanity*. Harcourt Brace.
Mumford, L. (1967/1970). *The myth of the machine. Vol 1, Technics and human development. Vol. 2, The pentagon of power*. Harcourt Brace Jovanovich.
Ortega y Gasset, J. (1914). Meditaciones del Quijote. : Publicaciones de la Residentia de Esudiantes.
Ortega y Gasset, J. (1939). Meditación de la técnica. In *Ensimismamiento y alteration*.: Espasa-Calpe. In *Obras Completas* (pp. 317–375). Madrid: Revista de Occidente 1945–1947.
Ruin, H. (2012). Technology as Destiny in Cassirer and Heidegger: Continuing the Davos Debate. In A. S. Hoed & I. Folkvord (Eds.), *Ernst Cassirer on form and technology: Contemporary readings* (pp. 113–138). Palgrave Macmillan.
Spengler, O. (1931). *Der Mensch und die Technik: Beitrag zu einer Philosophie des Lebens*. C.H. Beck.
Stiegler, B. (1998). *Technics and time, 1: The fault of epimetheus*. (trans. R. Beardsworth & G. Collins). Stanford University Press.
Stiegler, B. (1994/1996/2001). *La technique et le temps I, II et III*. : Galilée.
Stiegler, B. (2004). *Philosopher par accident*. Galilée.
Verbeek, P.-P. (2005). *What things do*. (trans. R.P. Crease). Pennsylvania State University Press.
World Transhumanist Association (1998). Retrieved from http://transhumanism.org. Accessed 21 Sept 2015.

Chapter 3
Derrida and the Phenomenology of Technology

3.1 Introduction

It is hard to see a clear philosophical direction in Jacques Derrida's writings. He has been described as a phenomenologist, a post-phenomenologist, a postmodernist, a post-structuralist, a philosopher of language, and even a relativist and a nihilist. He was long regarded principally as a philosopher of language or deconstructor of other philosophical and literary texts. Yet closer examination of his writings shows that he did not confine himself to deconstructive analyses of language and text. Aesthetics and politics were always important topics for him, even if he only felt able and willing to speak openly about them after the fall of the Berlin Wall.[1] In his later years he also took a growing interest in problems within the philosophy of religion.

Derrida's oeuvre of almost a hundred published works also contains countless instances of his philosophical reflections upon *technology*. Yet the aim of this chapter is not to show that Derrida is a philosopher of technology (though I argue he is that) but to try to illuminate *the way in which* technology and our relation to technology figure in his writings. In what sense can deconstruction be described as a meditation on technology? What does Derrida mean by technology? And what role does *writing* play in his philosophy of technology?

I begin by giving some brief background to Derrida's engagement with technology. Using his deconstructive analyses of Edmund Husserl's phenomenology, I show how Derrida gradually comes to view the question of *writing* as the central issue within the philosophy of technology. I then offer an account of his understanding of the concept of technology, followed by a discussion of Derrida's deconstruction of Martin Heidegger's philosophy of technology and the ways in which it differs from his own. I conclude by trying to summarize my findings.

[1] The reasons for Derrida's relative silence on political issues prior to the fall of the Berlin Wall are considered in Chap. 7.

B. Sjöstrand, *Derrida and Technology: Life, Politics, and Religion*,
https://doi.org/10.1007/978-3-030-83407-4_3

3.2 Derrida's Engagement with the Question of Technology

From the very start, Derrida was deeply engaged with the question of technology and our relation to technology. As we have seen, modern technology's spectacular advances served as a source of inspiration in his frequent discussions of genetic engineering, cybernetics, computer technology, archival technology, media technology, and telecommunication in works like *De la grammatologie, L'écriture et la différence, Mal d'archive, Échographies de la télévision, Foi et savoir, La carte postal*, *Papier Machine*, and *Negotiations* to name just a few of the most important. In many of his writings he is also deeply engaged in the question of how politics and democracy are affected by developments in technology (for example, in *Politiques de l'amitié, Spectres de Marx, Voyous*, and "Nietzsche and the Machine"). For Derrida, the concept of democracy, which has always been associated with the city, the state, the *polis*, and, in the modern era, the nation state, is confronting a powerful challenge in the limitless technology and media of today and tomorrow. To take an example, in the future new communication technologies may dramatically accelerate an ongoing global process of democratization. Derrida argues that the development of telephones, fax machines, televisions, email, and the Internet may even come to have a greater significance for democratization than the endless discussions of human rights and increased democratization.[2] There can be no question of halting this process of technologization; rather, it is a matter of understanding how a democratic response to this process might be formulated. Even though technology represents both terrifying dangers and fantastic possibilities, Derrida *welcomes* both technology and democracy.[3] He views technology itself as neither good or bad but *both* good and bad. The issue, as he sees it, is not about condemning or accusing technology, since nothing happens without technology; rather, it is a question of understanding what is happening and what will come about through and with technology.[4] Derrida is thus neither technophobe nor technophile, and his deconstructive thinking cannot be understood on the basis of these categories. Nor does he ever oppose technology itself (for him, there is no technology with a capital "T"); instead, he differentiates between the opportunities that various technologies offer for advancing or retarding the process of democratization. Technology must therefore be constrained by the ethical-political; it cannot be separated from the notion of responsibility and justice.

Derrida was also deeply engaged with questions of religion. Although he considered himself an "atheist," his life was in many ways shaped by his Jewish, Muslim, and Christian background. In his many writings on religion, among which *Donner la mort* and *Foi et savoir* deserve special notice, he often discusses in depth the relation of religion to technology, science, and, not least, various kinds of media.

---

[2] Derrida & Stiegler, *Échographies de la télévision*, p. 82.

[3] Derrida, "Nietzsche and the Machine", pp. 247–249.

[4] Ibid., p. 245.

Derrida regarded as crucial the relationship between technology and *time*, above all the future, which he designated the "to come" (*l'avenir*). By "to come" he did not mean the future in its usual temporal sense but something that is always arriving, always approaching, always already in the future.[5] For him, this absolute or structural future represents a condition of possibility for changes in the present. In this regard he refers to a "to come" democracy that has the character of a *promise* and, like a messianic promise of a coming Messiah, is a tool by which the future can intervene and affect the present. This involves a relation to that which is entirely other, something he calls the *messianic* (in *Specters of Marx*, for example); the messianic can come at any time, no-one can see it coming, no-one is forewarned, anything can happen. A situation so fundamentally insecure requires *vigilance* and *openness* to an unknown future that can never be predicted. Since such openness to the future can also be experienced as a threat, the "to come"—new technology, for instance—must be *interpreted*. It is necessary to *interpret in every situation* what one is doing and what one wants to do with technology.[6] In this regard Derrida is unambiguously a hermeneutically-oriented thinker. But, as I have already pointed out, his thinking is based not only upon Heidegger's hermeneutics but, above all, upon a lifelong dialogue with the writings of Edmund Husserl.

3.3 Husserl and the Technological Origin of Geometry

Technology is a highly visible presence in Derrida's first substantial work, *Le problème de la genèse dans la philosophie de Husserl*, which he originally wrote as a dissertation for his *diplôme d'études supérieures* in 1953. At the same time as he gives a highly informed overview of Husserlian phenomenology, he identifies a fundamental problem in it: right from the start, the problem with Husserl's thinking lies in the tension between explaining the genesis of ideal objects and preserving its *transcendental* character (as ideal objects, they are independent of historical-empirical experience). Although the "immense difficulty" with this paradox, which Derrida calls "the transcendental genesis,"[7] is a theme of several of Husserl's most important writings, it was not until his famous appendix to *Krisis*, "The Origin of Geometry" (1936), that Husserl offered a solution to the problem. In this classic short text, less than thirty pages long, he shows how geometry was able to emerge from more empirical and material practices.

However, from Husserl's perspective, the origin of geometry is not a historical-empirical question; rather than being about searching for the earliest geometricians

[5] According to Derrida, the future in its usual temporal sense can only "reproduce the present. Justice remains *to come*, it remains *by coming* [*la justice reste à venir*], it *has* to come [*elle a à venir*], it *is* to-come, the to-come [*elle* est *à-venir*], it deploys the very dimension of events irreducibly to come." (*Force of Law*, p. 27).

[6] Derrida, "Nietzsche and the Machine", p. 245.

[7] Derrida, *The Problem of Genesis*, p. xxxvi.

and their scientific theories, it involves an inquiry into the "most originary sense" and how that sense has been handed down the millennia all the way to the present.[8]

According to Husserl, geometry's original beginning, its "originary foundation," can only have been based, not on an objective scientific ideal, but on a pre-scientific lifestyle that was originally naïve. This naïve life, prior to knowledge, is what Husserl calls *lifeworld*. By gradually improving particular preferred shapes (lines, edges, surfaces, etc.), geometry was eventually able to detach itself from this empirical and material lifeworld and appear as an ideal form of knowledge, liberated from its material prehistory.

In purely concrete terms, Husserl images this future geometrician as first appearing in the form of a carpenter who, in response to a practical need, such as for an item of furniture, selects particular materials, such as planks, on the basis of a number of preferred forms: points, straight lines, right-angled corners, and flat surfaces. As these are gradually improved, pure geometric forms are gradually able to emerge:

> It is clear … that in the life of practical needs certain particularizations of shape stood out and that a technical praxis always [aimed at] the production of particular preferred shapes and the improvement of them according to certain directions of gradualness. First to be singled out from the thing-shapes are surfaces—more or less "smooth," more or less perfect surfaces; edges, more or less rough or fairly "even"; in other words, more or less pure lines, angles, more or less perfect points; then, again, among the lines, for example, straight lines are especially preferred, and among the surfaces the even surfaces; for example, for practical purposes boards limited by even surfaces, straight lines, and points are preferred, whereas totally or partially curved surfaces are undesirable for many practical interests. Thus the production of even surfaces and their perfection (polishing) always plays a role in praxis.[9]

What interests Husserl is how practical-technical activity gives rise to the geometric idea. Despite Husserl's idealism, this derivation clearly shows the close connection between scientific theory and pre-scientific, practically-oriented experience, insofar as science is always based on some form of technological praxis.

For the young Derrida, this conclusion, indeed, all of "The Origin of Geometry," is "most disappointing."[10] Husserl derives ideal objectivity by means of an unacceptable, in fact risible, "technological explanation"—a mistake that seems to obliterate everything that phenomenology has always uncompromisingly stood for. For Derrida, it seems as though transcendental intentional analysis has descended into "prephilosophical empiricism."[11] The *technological* nature of this explanatory hypothesis is clear from the fact that the origin of the ideal presupposes a "rigorous "measurability" [that] is born from the world of spatio-temporal things" and that

> its origin in human activity is purely technical; it is a "polishing" that gave us the pure idea of surface; out of lines and points that are "more or less pure" that geometric lines and

[8] Husserl, "The Origin of Geometry", p. 158.

[9] Ibid., p. 178.

[10] Derrida, *The Problem of Genesis* p. 166.

[11] Ibid., p. 167.

> points appeared. Again, the empirical technical and psychological act of "comparison" gave birth to identity. All the details of this curious analysis describe a purely technical genesis.[12]

The "purely technological origin" of ideal objects means that technology is inscribed within the genesis of the transcendental. The Derrida of 1953 is quick to distance himself from this position. But by the time he published his second book *Introduction à 'L'Origine de la géometrie' de Husserl* (1962; *Edmund Husserl's "Origin of Geometry": An Introduction*), he had reconsidered the issue entirely, endorsing the very position from which he had previously dissociated himself. Henceforth, technology as the origin of knowledge and meaning would become a central theme in all of Derrida's writings.

3.4 Husserl and Writing as Telecommunication

In his second book, Derrida returns to "The Origin of Geometry." However, this time the central issue is *writing*. In this key work, he shows that, for Husserl, writing is intimately associated with scientific objectivity, which he regards as the final and highest form of knowledge. It is not merely that scientific meaning ultimately becomes objective by being written down. In written form, meaning can also be preserved in a way that is unquestionably objective, being conveyed from generation to generation and incorporated into a common fund of experience, which future scientists can draw from and build upon.[13] Husserl writes in "The Origin of Geometry" that the enormous and complex theories which have been developed over the centuries would not have been possible without writing's documentary and preservational function.[14]

Put differently: for Husserl, writing is the very *condition of possibility* for all thinking and all knowledge. It is precisely the "iterative" nature of writing—its capacity to preserve and spread knowledge in various ways between vastly different cultural-historical epochs and settings—that accounts for both the possibility of a science that seeks objective truth and the possibility of a history in which we can identify the different stages in the development of science to the present day. Writing and written communication thus allow us to ask questions of tradition, "return inquiries" that Derrida likens to "postal and epistolary contact at a distance." For Derrida, writing makes possible "openness to a telecommunication."[15]

Why does Derrida place such emphasis on writing and written communication, rather than speech or the spoken word? Why does he upend the entire history of philosophy, which has consistently made writing subordinate to speech, whose absolute presence is taken as representing truth. The answer is simple. As he

[12] Ibid., pp. 168–169.

[13] Derrida, *Edmund Husserl's Origin of Geometry*, pp. 87–88.

[14] Husserl, "The Origin of Geometry", pp. 164–169.

[15] Derrida, *Edmund Husserl's Origin of Geometry*, p. 50.

explains, writing is: "a technique or a technical *means of communication.* Indeed, one is compelled to regard it as an especially potent means of communication, *extending* enormously, if not infinitely, the domain of oral and gestural communication."[16] Verbal communication is simply not sufficient to guarantee the existence of ideal signification. This is especially true of those instances in which the first geometricians are, for various reasons, *absent* or perhaps even dead.[17] In other words, there is an existential lack in the speaking subject's memory, which, being alive, is transitory. Writing here functions as a technological prosthesis, an *externalization* of memory, whose contents can be preserved in a way that is unquestionably objective and passed down to future generations.

Derrida goes a step further, however. Writing is not *only* a "worldly and mnemotechnological aid" to the subject, it is the very precondition for that subject being able to lay claim to objective truths in the first place. Since this precondition can hardly be empirical in nature, he raises the notion of a transcendental *epoché*, the idea of "a subjectless transcendental field." Writing represents such a transcendental field.[18] Accordingly, writing must appear on two separate levels: an *empirical* level that sets the technological-material (and "spiritual") preconditions for our *access* to knowledge; and a *transcendental* level that sets the preconditions for knowledge and meaning.

A key question of principle in this regard relates to our understanding of writing as *technology*: what is the relationship between empirical understanding of technology and the position advocated by Bernard Stiegler, namely, that technology also represents a transcendental, or, rather, quasi-transcendental, structure?[19] Stiegler

---

16 Derrida, *Limited Inc.*, p. 3.

17 Derrida, *Edmund Husserl's Origin of Geometry*, p. 87.

18 Ibid., p. 88.

19 Derrida introduced the phrase "quasi-transcendental" in *Glas* (1974). The first *explicit* sign of a modification of the classical transcendental structure is to be found in the first section of *De la grammatologie* (1965), where Derrida introduces the concept of "ultra-transcendental" in order to underscore the "necessity of a pathway (*parcours*)" through the transcendental problematic (p. 90). He stopped using the term after the publication of *Margue - de la philosophie* (1972). Instead, he began using "quasi-transcendental", which quickly became the dominant concept in his work. In broad terms, the prefix "quasi" means that deconstruction is a matter less of looking for a *deeper* transcendental origin (as the prefix "ultra" undeniably implies) than of showing that a given transcendental structure is always already contaminated by *the empirical* (the historical, the worldly, the temporal, etc). More specifically, the concept of "quasi-transcendental" expresses the idea that, beyond the notion of an absolute opposition between the empirical and the transcendental, the "condition of possibility" for any given structure is always already contaminated by the threat of its own impossibility or extinction. Although this formal law—the condition of possibility as a condition of impossibility—was only introduced *as such* after Rodolphe Gasché formulated it in *The Tain of the Mirror* (1986), it has been quietly at work in Derrida's writings from the beginning. Indeed, a good example is *Edmund Husserl's Origin of Geometry*, in which writing has a kind of quasi-transcendental status. The whole point of Derrida's argument there is that Husserl's demonstration of writing's transcendental role in the constitution of ideal objects is only valid if we regard writing as an empirical fact whose possibility is always necessarily threatened by the possibility of its own extinction. In other words, writing functions *simultaneously* as a condition of both possibility and impossibility—which is to say, as a quasi-transcendental.

formulates the question as follows: if things are as Derrida says, i.e. that a document's or a book's "volume and duration are neither purely sensible phenomena, nor purely intelligible noumena," does it not then become necessary to call into question the "ontological" status of *every* technological apparatus? Does technology not then become *constitutive*?[20]

The question of technology's status, that is to say, the relationship between empirical and quasi-transcendental understandings of technology, would be examined much later in *Échographies de la télévision* (1996), a book that records in concentrated form a conversation between Derrida and Stiegler. Thus, Derrida replies to Stiegler by clarifying the structure of topology:

> The origin of sense makes no sense. This is not a negative or nihilistic statement. That which bears intelligibility, that which increases intelligibility, is not intelligible—by definition, by virtue of its topological structure. From this standpoint, technics is not intelligible. This does not mean that it is a source of irrationality, that it is irrational or that it is obscure. It only means that it does not belong, by definition, by virtue of its situation, to the field of what it makes possible. Hence, a machine is, in essence, not intelligible.[21]

Derrida argues that it is both logically and topologically problematic to claim, as Stiegler does, that we can have knowledge of something (in this case technology) if that something is itself a condition of possibility. Stiegler's unconventional interpretation of technology has been criticized by Geoffrey Bennington and Ben Roberts, who argue that Stiegler has unacceptably conflated an empirical and a quasi-transcendental understanding of technology.[22] For Derrida, this is indefensible: quite simply, it is topologically impossible to reduce the quasi-transcendental level to technology.

To summarize: for Derrida, *writing* represents the ultimate guarantor not only of the objectivity of scientific meaning but also of its genesis and transmission in time and space. As such, writing is a technological tool that makes possible the opening of a telecommunication through which "reconnective" questions can be asked of the scientific tradition. Derrida's privileging of writing, which involves an inversion of the traditional speech/writing hierarchy, necessarily requires us to understand writing in two ways: writing in its usual empirical sense (as, for example, the writing on this page); and writing as a subjectless condition of possibility for knowledge and meaning—a quasi-transcendental that Derrida will later rename "arche-writing" (*archi-écriture*).

---

[20] Stiegler, "Derrida and Technology", p. 247

[21] Derrida & Stiegler, *Echographies de la télévision,* p. 121. On the same page, Derrida adds that the lack of meaning in technology can leave us depressed and result in dehumanization and nihilism. Or, as he rather more forcefully puts it: "in itself, this non-sense is not an absurdity, it is not negative, but it is not positive either."

[22] Bennington, "Emergencies", p. 190 and Ben Roberts, "Stiegler reading Derrida", p. 14.

3.5 Technology as a Relation

It is important to understand that Derrida did not formulate his philosophy of technology out of thin air but took as his starting point Husserl's ideas about writing in "The Origin of Geometry." These ideas are a source of inspiration for Derrida when he develops his concept of arche-writing. Yet the question posed in "The Origin of Geometry" heralds not only arche-writing but also other quasi-transcendentals such as "trace" and *différance*. In *Speech and Phenomena*, Derrida describes how a "trace" of the past not only opens up the pure presence of the present and, with it, the history of life and consciousness, it also constitutes it by virtue of the "movement of *différance*" that it achieves.[23] Derrida describes a network of relationships, a "web" of traces, with no stabilizing centre, no pure presence. This quasi-transcendental structure—if such a formulation can be allowed—is "arche-writing."[24]

The structure that emerges here rests upon an entirely new logic, *the logic of the supplement*, which Derrida develops in the context of a reading of Rousseau's writings.[25] The logic of the supplement differs from the logic of the opposite, according to which one term in a binary opposition serves as an externally-derived supplement to the other: culture is a supplement that derives from nature, technology is a supplement that derives from culture, and so forth. In the logic of the supplement, by contrast, this externality is a necessary complement of the first term, supplementing its lack so that it can appear in the first place. Thus, if there can be no nature without culture, and no culture without technology, then there can be no nature without technology. For deconstruction, nature and technology, *physis* and *techne*, are not each other's opposite but always already inescapably intertwined. Technology is *a priori* a necessary supplement to nature:

> Thus one could reconsider all the pairs of opposites on which philosophy is constructed and on which our discourse lives, not in order to see opposites erase itself but to see what indicates that each of the terms must appear as the *différance* of the other, as the other different and deferred in economy of the same (…culture as nature different and deferred, differing-deferring; all the others of *physis—techne, nomos, thesis*, society, freedom, history, mind, etc.-as *physis* different and deferred, or as *physis* differing and deferring. *Physis* in *différance*.[26]

Technology, nature, culture, society, consciousness, and so forth are all closely interwoven with each other as *différance*. Deconstruction therefore automatically regards as problematic any attempt to oppose, for example, something technological to something non-technological.

What are the implications of this new structure and this new logic for the *question of technology*? How does deconstruction view technology? Is technology

---

[23] Derrida, *Speech and Phenomena*, p. 134.

[24] Derrida, *Marges de la philosophie*, p. 13–17.

[25] Derrida, *De la grammatologie*, pp. 203–234.

[26] Derrida, *Marges – de la philosophie*, p. 18.

"merely" the sanded planks created by the first geometrician? Or, more generally, is technology just a material supplement to what Freud called "the psychic apparatus"? No, says Derrida, we can no longer talk about an opposition between the psychic apparatus (that is life) and the non-psychic apparatus (that is death). In reality, these two apparatuses *resemble* each other. And that resemblance is growing. The analogy between today's complex systems for the inscription and storage of almost unlimited quantities of information in exterior "memories," and our own limited and flawed memories, illustrates the need to reconsider the question of technology:

> Here the question of *technology* (a new name must perhaps be found in order to remove it from its traditional problematic) may not be derived from an assumed opposition between the psychical and the nonpsychical. Life and death. Writing, here, is *techne* as the relation between life and death, between present and representation, between the two apparatuses. It opens up the question of technics: of the apparatus in general and the analogy between the psychical apparatus and the nonphysical apparatus.[27]

When Derrida refers to technology as the relation between the psychic and the non-psychic, the inner and the outer, he normally uses the term *la technique*. For the non-psychic, he variously refers to technology, machine, or apparatus (*technologie, machine, appareil*). For instance, he speaks of new technologies or tele-technologies and, above all, of the teletechnological apparatus or machine as an umbrella heading for the press, radio, TV, internet, cellphone network, digital archives, etc.[28]

Deconstruction does not consider technology as an opposite term but as a *relation* between the psychic and the non-psychic, between people's living memories and external "memories," between the inner and the outer, insofar as the inner represents a condition of possibility for the outer even as the outer represents a necessary supplement to the inner if the latter is even to function in the first place. Technology *as* writing is thereby opened up in the difference between the inner and the outer, between what Plato called *mneme* and *hypomnesis*, where the distance separating them is even thinner than a hair: it is quite impossible to tell them apart.[29] With time, this view of technology as such a close relation came to shape Derrida's attitude towards Heidegger and his question of technology.

---

[27] Derrida, "Freud et la scène de l'écriture", p. 337.

[28] A word of clarification is needed with regard to the terms "technology","technics", and "the technical". The French term *la technologie* is generally translated as "technology", while *la technique* has a broader designation and is usually translated as "technics" or "the technical". For stylistic reasons, I will use both interchangeably.

[29] In "La pharmacie de Platon" (*La dissémination*, pp. 77–214), Derrida shows that Plato distinguishes between two kinds of memory: *mneme* and *hypomnesis*. The first, which lies in the human soul, is active and living and is characterized by the capacity for philosophical reflection, which, according to Plato, is the hallmark of all true knowledge. The second is an outer "memory", a text, stored in a substratum or supplement (such as a piece of paper, a phone, or a computer). For Plato, this outer memory is secondary, technical, and dead. For Plato, *mneme* and *hypomnesis* represent a hierarchical binary opposition in which *mneme* is privileged and *hypomnesis* is devalued. For Derrida, this hierarchical opposition is the very archetype of all metaphysical thinking and is the driving force of an intense desire to eliminate the technological from the human soul.

3.6 Heidegger and the Question of Technology

Martin Heidegger has had an immense influence on Derrida's thinking and is everywhere present in his writings. Derrida was also one of the first French philosophers to acquire a thorough knowledge of Heidegger's later work. It is hard to disagree with his view that Heidegger was one of the first thinkers to address in a rigorously philosophical way the big question of what technology really *is*.[30] Indeed, the question of technology is a fundamental theme in particular of Heidegger's later work, written in the 1950s and after. With growing insistency, Heidegger asks questions about modern technology and the modern lifestyle characterized by unchecked consumption, superficial values, and the plundering of nature with which it was associated. Heidegger's writings undeniably exhibit elements of a critique of technology as well as a belief that the era of modern technology must be passed through before something new and better can emerge.

For Derrida, Heidegger's thoughts on technology seem ambivalent: "There is no simple evaluation of technics in Heidegger."[31] Yet this did not stop him from trying to distance himself from his mentor by foregrounding a number of "difficulties." One of these difficulties involves the relationship between *physis* and *techne*. According to Derrida, Heidegger has a tendency of downgrading *techne* to something secondary in relation to a pre-technical originariness, *physis*. Clearly, this *physis* is not what we customarily refer to as "nature," but insofar as *physis is*, insofar as *physis* is existence itself in its entirety, it is superior to *techne*. For Heidegger, there appears to be something in *physis* that resembles a truth independent of *techne*. As Derrida asks rhetorically: "How can we overlook a Heideggerian "pathos" which, despite so many denials on this subject, remains antitechnological."[32]

Another difficulty concerns the relationship between technology and the *essence* of technology. Although Heidegger often discusses questions of a technical nature, what matters to him is *not* describing technology itself correctly. Instead, he wishes to approach the question of technology by considering its essence. In one of his best-known and most influential essays, *Die Frage nach der Technik* (1954, translated as *The Question Concerning Technology*), he famously argues that "The essence of technology is by no means anything technological."[33] For Derrida, this statement is at least in part a genuinely philosophical statement. It insists upon the possibility of a mode of thought that poses questions, fundamental questions about the essence of technology, from a position outside of technology itself. Only by means of such a mode of thought, detached from our involvement in technology itself, it is possible to understand what technology *is* and to formulate a satisfactory answer to the problems that modern technology has created.

---

[30] See, for example, *Écographies de la télévision*, p. 148 and "Nietzsche and the Machine", p. 244.

[31] Derrida, "Nietzsche and the Machine", pp. 244–245.

[32] Derrida & Stiegler, *Échographies de la télévision*, p. 149.

[33] Heidegger, *The Question Concerning Technology*, p. 4.

Although Derrida to some degree recognizes the merits of this approach, he sees a pressing need to deconstruct this "desire for rigorous non-contamination."[34] As we have seen, deconstruction relies on the logic of the supplement, according to which inner and outer are always already interwoven. In Heidegger, by contrast, there is a sharp line of demarcation between essence and supplement, between the essence of technology and technology itself, which Derrida regards as an echo of the ontological difference (according to which the essence of being is distinct from being itself).[35]

Derrida asks why Heidegger wants to "purify" thought of technology.[36] Whence this desire for "purification"? What does this say about Heidegger's view of thought as a whole? The answers to these "immense questions," Derrida writes, are anything but "self-evident." Instead of "purifying" thought, however, he insists upon

> the necessity, one could say the fatal necessity of a *contamination*—and the word was important to me—of a contact originarily impurifying thought or speech by technology. Contamination, then, of the thought of essence by technology, and so contamination by technology of the thinkable essence of technology—and even of a question of technology by technology, the privilege of the question having some relation always already, with the irreducibility of technology. It is to imagine that the consequences of that necessity cannot be limited.[37]

For Derrida, not only essentialist thinking but even Heidegger's privileged question—*Die Frage nach der Technik*—is contamined by technology itself. This is true of human life and life in general. Life, from the "genetic inscription" that regulates "the behaviour of the amoeba or annelid" all the way up to *Homo sapiens*, is always already contaminated with technology. From its earliest beginnings and in every moment, life applies a prosthetic strategy: life is a process of self-repair and its reproduction is a *mechanism*, a form of technology. It is not simply that technology is not the opposite of life, it is inscribed into life from the very beginning.[38]

Contamination, the "infection" that arises from a particular kind of contact, renders impossible all attempts to bracket essentialist thinking from technology itself. The question is whether it thereby represents a threat to the central term in Heidegger's thinking—ontological difference. For Derrida, the difference between Being and being seems to arise *within* being, which is thereby considered a form of contamination. This contamination demands a mode of thinking in terms of *différance* that is not the same as ontological difference.

Heidegger himself would perhaps have objected to the requirement of contamination that Derrida had not fully grasped the central point in his thinking—essentialist thinking and ontological difference. He might well have added—surely unjustly—that Derrida's "law" of contamination is only an ontic rule, a pure metaphor.

---

[34] Derrida, *De l'espritt*, p. 26.

[35] Derrida, *Échographies de la télévision*, p. 149.

[36] Ibid., pp. 133–4. The terms "pure", "purification", "and "purifying" are Derrida's own, not Heidegger's.

[37] Derrida, *De l'esprit*, p. 26. See also his "Nietzsche and the Machine", p. 236 in which he claims that contamination is "inevitable" and "absolutely undeniable".

[38] Derrida, *De la grammatologie*, p. 126.

A third difficulty concerns the relationship between the spiritual and the technological-material. For Heidegger, human life is intimately bound up with "the world." In his controversial *Rectorship Address* (1933), he not only asks the question of what the world *is*, he also gives the answer: "The world is always a *spiritual* world." This means that human life is fundamentally spiritual (*geistich*).[39] Surprisingly, Heidegger uses the German word *Geist*, meaning both spirit and spiritual being, despite expressly stating in *Being and Time* that the term should be avoided.[40] Derrida interprets this metaphorical gesture as a desire on Heidegger's part to save the (Western) subject from the loss of its unique position of power in the world. What saves humanity is precisely this (non-Christian) *Geist*, "*Geist*, whose purity Heidegger wants to save, a purity internal to spirit."[41] Nonetheless, the fact that Derrida calls into question this "pure" form of spirituality, *Geistigkeit*, does not mean that he would go so far as denying the existence of "the spiritual" in the world; the spiritual is, rather, always already contaminated by the technological-material.

Despite these three difficulties—the relationships between (1) *techne* and *physis*, (2) the essence of technology and technology itself, and (3) the spiritual and the technological-material, Derrida can hardly be said to disassociate himself from what Heidegger says. He merely calls into question Heidegger's anti-technological tendencies and his claim that such a thing as the essence of technology or pure spirituality can really exist independently of the technological-material.

The fact that Derrida calls into question the existence of an essence of technology independent of technology itself does not mean that he and Heidegger approach the question of technology differently. For Heidegger, as we have seen, the most important thing is not to discuss technology *as such*, without reflecting upon its *essence*. To take a topical example, the debate over nuclear power is, for Heidegger, merely the expression of an existential crisis that has its basis in our loss of Being's original meaning. This loss is the result of two thousand years of egoistical thinking, which have entirely obscured our true relation to being. In the process, the original meaning of *techne* has also been lost. With the advent of modernity, we begin to understand *techne* in the limited sense of "technology." Our understanding of modern technology, Heidegger argues, has come to be associated with the manipulation of nature through the application of scientific knowledge to large-scale industrial production. For modern human beings, the world is a resource for technological

---

[39] Derrida, *De l'esprit*, p. 75. As early as *Being and Time*, Heidegger claims that *Dasein* is spiritual (*geistig*) (see §70). In fact, he had examined the Pauline concept of spirit, *pneuma*, as far back as 1921, in lectures on the phenomenology of religion that were only published in 1994. This means that these early texts were unavailable to Derrida when he treated this subject in *De l'esprit*, which was published in 1987. Heidegger also returns to the concept of spirit in later texts such as *Introduction to Metaphysics* and in his readings of Hölderlin and Trakl. See Ruin, "Anxious Spirits: Pneumatology in Heidegger, Paul and Kierkegaard", pp. 4–5.

[40] Derrida, *De l'esprit*, p. 11.

[41] Ibid. p. 26. On Derrida's reading of *spirit* in Heidegger's inaugural lecture as vice-chancellor of Freiburg University, see also pp. 63–5.

manipulation within the framework of its usefulness. This framework or "enframing" (*Ge-stell*) is, as we have seen, the *essence* of modern technology.

In our technological culture, humanity is no longer the master of technology; instead, technology itself represents humanity as a framework or enframing, within which everything is interpreted. Humanity itself is "enframed" within technology. The framework is the source of modern humanity's fundamental attitude, which not only provides the basis for the development of modern technology but also permeates our perception of virtually everything in the world. This means that the world "is" always already determined when humanity makes its entrance—"Being" already has a meaning. This meaning is the basis for humanity's relation to the world; it determines how the world "is" for humanity. The framework is humanity's destiny, and no further questions can be asked about the way in which Being appears. Consequently, the framework—as the essence of technology—is synonymous with *Being itself*.

In contrast to Heidegger, Derrida is primarily interested not in the essence of technology but in our relation to different *technologies*. As we have seen, he defined technology in terms of relationships between different psychic processes and different technologies for the inscription and storage of all text, from clay tablets and papyrus rolls to books, e-books, digital archives, and the World Wide Web. Deconstructive analysis engages with the potential of these technologies to promote or hinder the development of democracy.

As an example, let us consider the way in which Derrida analyses "the archive," a central concept in his philosophy of technology and one that we will be coming back to on several occasions in this study. What is an archive? In *Mal d'archive* (1995), Derrida deconstructs the concepts of *archive* and *archiving* (in the sense of collecting, inscribing, and storing text). He begins by making an archival search of the term *archive* and its etymological origin, and discovers that the Greek word *arkhe* means both beginnings (*commencement*) and command (*commandment*). The archive is a place both where something has its beginnings—physically, historically, or ontologically—and a place where gods or people issue *commands* on the basis of a law or record. For Derrida, the archive is both the record itself and the place where the record is preserved. Accordingly, the archive becomes the place from which political power is exercised. In reality, all archives are closely bound up with the political power and the violence that both establish and conserve the law. The archive is both revolutionary and conservative.[42]

The archetypical archive contains written documents that relate to the public and that in some political systems are available to citizens. Yet citizens have *limited* access to these documents because the facility in which they are stored—the *archeion*—is both a private building *and* a public space controlled by its owners, the *archontes*—the principal rulers and founders of the law. The archive is thus strongly associated with a judicial order that encompasses things like family law, constitutional law, and the relationship between secret and not secret, private and public,

[42] Derrida, *Mal d'archive*, pp. 11–20.

with the latter being regulated by laws relating to ownership, usufruct, copyright, and so on. For Derrida, the question of a "politics of the archive" is therefore a privileged political question:

> This question will never be determined as one political question among others. It runs through the whole of the field and in truth determines politics from top to bottom as *res publica*. There is no political power without control of the archive, if not of memory. Effective democratization can always be measured by this essential criterion: the participation in and the access to the archive, its constitution, and its interpretation.[43]

The more democratic a society is, the more its archives and collective memory are controlled by the public, its citizens. And conversely: the less democratic a society is, the more its archives are controlled by political authorities who want to limit or completely prevent the public's access to archives. The question of the technological archive is, in other words, a fundamentally *political* question whose answer cannot be simply articulated and archived, regardless of how legitimate this question seems to be. The technological archive does not allow itself to be determined in an objective and neutral way. We can and should inquire into the meaning of this concept by referring to the past, collective memory or tradition. For Derrida, the question of the archive is not a question about the past. It is a question about the future, a question about the future itself, about the *promise* of a democracy to come.

3.7 Summary

Derrida is well versed in the philosophical tradition that tries to understand technology from a hermenutic-phenomenological perspective. At the same time, his thinking ranges across an unusually broad spectrum: in addition to fundamental philosophical reflections on technology, it includes the relation of technology to humanity, nature, time, ethics, politics, religion, and science. From his very first writings and all through his entire oeuvre, technology is a highly visible presence, not only in the form of technical terminology and metaphors but above all in his acute discussions of writing, teletechnology, and every form of media, from the origins of life, through the emergence of human beings, and right up to the advanced prosthetic strategies of our own era: the internet, the World Wide Web, email, archival technology, recording technology, biotechnology, and so on. The philosophical basis for these discussions is primarily set out in his analyses of the work of Husserl and Heidegger.

I have sought to give an introduction to Derrida's philosophy of technology by discussing his deconstructive analyses of Husserl and Heidegger. I am convinced that these analyses can make a valuable contribution to phenomenology and the philosophy of technology. Perhaps we ought to refer to a *phenomenology of*

[43] Ibid., p. 15.

technology in which *writing* in its general sense, rather than the transcendenal subject, is the fundamental concept.

What makes Derrida's thinking unique is surely his broadening of the concept of technology to include the entire field of phenomenology: there is no way that essentialist thinking can be bracketed off from technology (that is to say, the non-essence aspect of technology). The very desire to shut oneself off from technology makes us more vulnerable to the risk of being "infected" by it than if we try to think about it in terms of *contamination*. For Derrida, technology is not an opposition between essence and supplement but a close relation, a contamination, in which essence and supplement are intertwined. We must, he argues, continue this contamintion. Without it, we would be unable to take responsibility for technological development and globalization. Without it, we would be unable to remain open to that which is entirely other, to the messianic, and to the future with all its threats and opportunities.

References

Bennington, G. (1996). Emergencies (Vol. 18, pp. 175–216). *Oxford Literary Review*.

Derrida, J. (2003). *The problem of genesis in Husserl's philosophy*. (trans. M. Hobson). The University of Chicago Press.

Derrida, J. (2002). Nietzsche and the machine., The aforementioned so-called human genome. In *Negotiations: Inventions and Interviews*, 1971–2001. Elisabeth Rottenberg (Ed. and trans).: Stanford University Press.

Derrida, J. (1995a). *Mal d'archive: Une impression freudienne*. Galilée.

Derrida, J. (1995b). *Archive fever: A Freudian impression*. (trans. E. Prenowitz). The University of Chicago Press.

Derrida, J. (1992). Force of law: The "Mystical Foundation of Authority". In D. Cornell, M. Rosenfeld, & D. G. Carlson (Eds.), *Deconstruction and the possibility of justice*. (trans. M Quaintance). Routledge.

Derrida, J. (1989a). *Of spirit: Heidegger and the question*. (trans. G. Bennington, R. Bowlby). The University of Chicago Press.

Derrida, J. (1989b). *Edmund Husserl's 'Origin of geometry': An introduction*. (trans. J.P. Leavey). University of Nebraska Press.

Derrida, J. (1988). In G. Graff (Ed.), *Limited Inc.*. (trans. S. Weber, J. Mehlman). Northwestern University Press.

Derrida, J. (1987). *De l'esprit: Heidegger et la question*. Galilée.

Derrida, J. (1982). *Margins of philosophy*. (trans. A. Bass). The University of Chicago Press.

Derrida, J. (1978). *Writing and difference*. (trans. A. Bass). The University of Chicago Press.

Derrida, J. (1974). *Glas*. Galilée.

Derrida, J. (1973). *Speech and phenomena, and other essays on Husserl's theory of signs*. (trans. D. B. Allison). Northwestern University Press.

Derrida, J. (1972a). *La dissemination*. Seuil.

Derrida, J. (1972b). *Marges – de la philosophie*. Minuit.

Derrida, J. (1967a). *De la grammatologie*. Minuit.

Derrida, J. (1967b). *L'écriture et la difference*. Seuil.

Derrida, J. (1967c). Freud et la scène de l'écriture. In *L'écriture et la difference* (pp. 293–340). Seuil.

Derrida, J., & Stiegler, B. (2002). *Echographies of television: Filmed interviews*. (trans. J. Bajorek). Polity Press.

Derrida, J., & Stiegler, B. (1996). *Échographies de la télévision*. Galilée.

Gasché, R. (1986). *The Tain of the Mirror: Derrida and the philosophy of reflection*. Harvard University Press.
Heidegger, M. (1977). *The question concerning technology and other essays*. (trans. W. Lovitt). Harper and Row.
Heidegger, M. (1962). *Being and time*. (trans. J. Macquarrie & E. Robinson). Harper and Row.
Husserl, E. (1978). The origin of geometry (trans. D. Carr). In J. Derrida (Ed.), *Edmund Husserl's origin of geometry* (pp. 155–180). University of Nebraska Press.
Roberts, B. (2007). Stiegler reading Derrida: The prosthesis of deconstruction in technics. *Postmodern Culture, 16*(7).
Ruin, H. (2014). Anxious spirits-pneumatology in Heidegger, Paul, and Kierkegaard. *Topos, 1*, 39–52.
Stiegler, B. (2001). Derrida and technology: Fidelity at the limits of deconstruction and the prosthesis of faith. In T. Cohen (Ed.), *Derrida and the humanities*. Cambridge University Press.

Chapter 4
Technology and Time as Event

Time is artificially produced. It is an artifact—Jacques Derrida

4.1 Introduction

Joining technology and time in a heading might seem like a stretch. What has technology got to do with time? How can something as visible as technology have anything in common with time, the most invisible dimension of our existence? Might we not go so far as to ask whether time even exists? And yet: we know from personal experience that for anything to *appear*, it needs and takes time. We know that no phenomenon sees the light of day other than during the day, that is say, during the time the sun travels across the heavens, rising in the east and setting in the west. We also know, or think we know, that everything goes much *more quickly* than before. Yet there can be few people who seriously doubt that the faster tempo and rapid changes in our rhythms of living are direct consequences of technological development. To take a topical example, handwritten letters that once took several days to reach other cities in Europe have now been replaced by *e-mails* that are distributed at lightning speed across the whole world.

Although Derrida is interested in questions relating to this chronological or cosmological time, what really engages him is another question: the question of our *experience* of time and how that experience is related to technology. This question is what we will be considering in this chapter. A number of follow-on questions present themselves. How does Derrida view Martin Heidegger's attempt in *Being and Time* to ground the experience of time in an originary, existential temporality? Conversely: how does he regard the perspective put forward by Bernard Stiegler in his monumental study of technology and time, *La technique et le temps* (translated as *Technics and Time*), namely that time is fundamentally technological? As I will try to show in this section, Derrida calls into question both these positions on the grounds that they try, in differing ways, to establish an ultimate foundation for the phenomenon of time. For Derrida, no such ultimate foundation exists. Time does

B. Sjöstrand, *Derrida and Technology: Life, Politics, and Religion*,
https://doi.org/10.1007/978-3-030-83407-4_4

not have an ultimate foundation, regardless of whether that foundation is called time or technology.

To understand how Derrida views the relation between time and technology, we first need to rehearse what he means by technology. As we saw in the previous chapter, the concept of technology does not refer solely to a technical apparatus. Such an apparatus never works by itself. It requires a relation to human life in order to operate, and without such a relation it is and will remain dead. To have life, the technical apparatus must be connected to life, to the living present, to the psychic in people, what Freud calls the psychic apparatus. The technical apparatus should therefore be understood as a necessary *supplement* to the psychic apparatus. The supplement is added to the psychic as a means of expanding its potentialities and in order to remedy its obvious failings and limitations. The question of technology can thus not be understood on the basis of an absolute *opposition* between the psychic and the non-psychic, between life and death. Rather, technology must be viewed "as the *relation* between life and death, between present and representation, between the two apparatuses."[1]

The fact that Derrida regards technology as the psychic apparatus's relation to the technical apparatus means that the technical apparatus is in some way connected with the phenomenon of time. To understand this relation, we therefore need to begin by examining what Derrida understands by time. Accordingly, I will begin this section with a summary of his analysis of what Heidegger called the "vulgar" sense of time, which both men regard as a hallmark of the philosophical tradition going back to Aristotle. My aim is not to survey the history of the philosophy of time but to show that there is a clear line running from Aristotle, via Hegel, to Derrida as regards their view of time as a *becoming-space*, an understanding that leads up to Derrida's notion of *différance* as the meaning-making framework for time. I then discuss Heidegger's strategy of contrasting "vulgar" time with an original time and Derrida's reasons for questioning this strategy even as he builds on Heidegger's other ideas about time, particularly those relating to time as a series of events. This discussion then moves on to Derrida's own conception of time, in which time consists precisely of a series of *events* that suddenly intervene and interrupt the homogenous flow of chronological time. As a result, time must be regarded as no longer linear but, rather, dislocated. As he summarizes his understanding of time, with a quotation from Hamlet's monologue: *The time is out of joint*.

I also summarize Derrida's view that events are increasingly being produced *artificially* by a global media apparatus comprised of the press, radio, TV, and, in our own time, social media. These "current" events are in fact artificial creations, which have to be differentiated from the original events themselves. Finally, I confront Derrida with Stiegler and the main thesis of the latter's *La technique et le temps*, namely that *all* time is technological time. There is, I propose, an irresolvable conflict between their respective understandings of the relation between technology and time.

---

[1] Derrida, *L'écriture et la différence*, p. 337.

4.2 The So-Called Vulgar Concept of Time

In a famous footnote, by far the longest in *Being and Time*, Heidegger analyses the origin and development of the so-called vulgar concept of time, from Aristotle to the present. This concept of time, which Heidegger coined in order to contrast with his own, is based on the assumption that time as such is something that can be measured, for example by clocks. More precisely, time is to be understood as a succession of *nows* that stretches from the future to the past. In one of Derrida's more central texts, *Ousia et Grammè*: *note sur une note de Sein und Zeit* (1972), he discusses Heidegger's analysis of vulgar time and the question of the now and presence. Derrida's starting point is that the dominance of the now has never been questioned by the philosophical tradition:

> Has not the entire history of philosophy been authorized by the "extraordinary right" of the present? Have not meaning, reason, and "good" sense been produced within this right? … How could one think Being and time *otherwise* than on the basis of the present, in the form of the present, to wit a certain *now* in *general* from which no *experience*, by definition, can ever depart?[2]

According to Derrida, the now has never seriously been questioned. This now may seem completely self-evident, something outside of which no thinking in the world should be possible. And in the entire history of metaphysics, meaning has never been understood upon any other basis than the now and presence as presence. Non-presence has always been thought about by means of a modalization of presence. The past and the present have always been thought about as past or present nows.[3] Despite its everyday obviousness, however, closer analysis of this *impossibility* of conceiving of time *in another way* leads to a number of contradictions, so-called *aporias*, which seem impossible to escape and which have preoccupied philosophers since antiquity.

The first to consider the aporias of time in systematic fashion was Aristotle. His truly fundamental investigation of time is outlined in the fourth book of the *Physics*. It is here that the word "aporia" appears for the first time, and it is here that Aristotle discusses the aporias linked to time. The aporias of time are also the subject of *Ousia et Grammè*, which additionally provides an impressive historical reconstruction of the metaphysics of space and time. In this text, Derrida shows that the philosophical tradition from Aristotle to Hegel has always regarded time as a series of now-points moving along *an imagined line*. The origin of this metaphysical linearity is the very text by Aristotle that Derrida painstakingly analyses.

According to Derrida, Aristotle was actually wrestling with two closely-related aporias that address the same problem: *time is what it is not*. The first aporia means that if time consists of a series of nows, it cannot be defined as either being or not-being: time "is" and "is" not what it is. The question now is, how should we understand the statement that time is something that it is not? Aristotle invokes our

[2] Derrida, "Ousia et Grammè", p. 41.

[3] Ibid., p. 37.

everyday experience, in which time nonetheless seems to exist in the form of a *now*. This now is the only way in which time can exist. And yet this now in some sense *does not exist*. This now is, in fact, a border between that which *no longer* is and that which is *not yet*. Or, as Aristotle puts it: "In one sense it has been and is no longer, and in another sense, it will be and is not yet" [217b]. In other words, time is comprised of *no-thing*, and as such it does not belong among things that have being. For this reason, time poses a challenge to our logical thinking, which continually alternates between two contradictory statements: *time is* versus *time is not*.

The second aporia proceeds from the first. If time is indeed comprised of a series of nows, what is it *then*? According to Aristotle, time is bound up with movement [219a]. Time exists in a relation to movement, but the question is: what does this relationship look like? Once again Aristotle refers to our everyday way of thinking. We ourselves know that time is always bound up with some kind of movement, for example, that of the Sun, the Earth, or the Moon. Because Aristotle, for good reasons, denies that time and movement are the same, he chooses to define time as *a number of movement following the before and the after* [219b]. The concepts of "before" and "after" refer to the fact that two different nows are always "before" or "after" in relation to each other, much the same way as when two railway carriages in a long train go past: one passes "before," the other "after." When we imagine the before and after of movement, the end-points of that movement correspond temporally to two different nows. Quite simply, time is what exists between these two different nows: "That which is marked off by the now is thought to be time" [219a]. Time is thus a *number* for movement, which paradoxically means that time is something mathematical, even though time in itself, by its nature, is not mathematical. Time as such is not a number, a figure in the series 1, 2, 3..., any more than horses and people as such are "the number that counts them." As a result, we need no longer conceive of time as a being among other beings, along with horses and people.

Defining time becomes even more problematic if we consider movement as a series of points that collectively form a line. Aristotle rejects this view, too, on the grounds that the points as such are not in motion but static; they constitute a series of potential *limits* that prevent time's *continual* flow. However, he admits that even though time cannot be reduced to a simple graphic representation, what he calls a *gramme*, it is impossible not to describe it as a line. Derrida summarizes the aporia of linear time as follows: "[I]t is because movement is determined according to the before and the after that the graphic linear representation of time is simultaneously required and excluded by Aristotle."

The solution to the problem of the point and the line, according to Derrida, is to think of the line not as a series of potential limits "but as a line in act, as a line thought on the basis of its extremities."[4] The line must be thought of as an active line along which every now, every point, is continually moving through space-time. A paradoxical consequence of this solution is that the line must be imagined as a *circle*. Because the two end-points of the line meet, the line must always be understood

[4] Ibid., p. 69.

within the framework of a "circle that regenerates itself *indefinitely*."[5] The image of time as a circle is, in fact, one of the most powerful images in metaphysics, something that Derrida already refers to in his second book, *Edmund Husserl's Origin of Geometry*, and that he returns to throughout his writings.[6]

Time must always already be understood spatially in terms of points, lines, and circles. It is, in other words, a *becoming-space*. In order to prove his thesis about the becoming-space of time, Derrida examines Hegel's philosophy of time and space, which he regards as a paraphrase of Book Four of the *Physics*.[7] Hegel's dialectical repetition of Aristotle's aporias is most clearly visible, he argues, in *Philosophy of Nature*, where time and space are treated as fundamental categories in nature. According to Hegel, nature is originally an indifferent, indeterminate, and continual space—"absolute space." In order to become nature in the usual sense, this indeterminate, indifferent space must be differentiated and defined. For Hegel, this occurs through the negation of the absolute space's original purity: the point, the line, and the plane are generated from this space by means of repeated negations: first, the point is produced by the negation of absolute space, then the line by negation of the point, and finally the plane by negation of the line. Hegel then argues that this dialectical process in space *is always already time*. At every step of the process *time is required*. For this reason time in itself is only a *becoming-space*, which is to say, space is time, it temporalizes itself, it realizes itself as time. We will shortly see how this *space-becoming* also reappears in Derrida's description of *différance*.

4.3 Heidegger's Time

It may seem strange that Derrida's *Ousia et Grammè* investigates Aristotle's and Heidegger's ideas about space-time and Heidegger's attempt to rethink time and being, rather than engaging with Husserl's phenomenological analysis of the concept of time, for example in his 1905 lectures on "inner time-consciousness." The reason is that Derrida had already examined Husserl's concept of time, particularly in *Speech and Phenomena* (1967), and that he was strongly influenced by Heidegger's analyses of time, particularly in *Being and Time* (1927).

The overall focus of Heidegger's magnum opus may be the issue of Being, but in practice it is about the existential structures that constitute human existence (*Dasein*). In Division Two of the volume, Heidegger shows that these structures can only be understood if they are interpreted temporally. Temporality is understood as a fundamental condition of possibility that serves as the basis for *care*, the fundamental existential structure of *Dasein*. It is primarily through care that the so-called

[5] Ibid.

[6] Derrida, *Edmund Husserl's Origin of Geometry*, pp. 124 and 133.

[7] Derrida, "Ousia et Grammè", pp. 44–51. Derrida further argues that Kant can be situated "in the *direct line* which … leads from Aristotle to Hegel" (p. 44).

"temporal ecstasies"—vectors thrown to the past, the present, and the future—form a unity which cannot be divided. Although "vulgar" time, that is, time measured by calendars and chronometers, dominates in the metaphysical tradition, it is not fundamental but depends on a more originary "temporal-making." The essence of time is not a series of nows. It is a fundamental condition of existence.

Even if an existential understanding of the phenomenon of time is productive in many ways, Heidegger only treats its main features in *Being and Time*. Moreover, in his later writings he moves away not only from the existential analysis of *Dasein* but from an understanding of time's relation to *Dasein* as a whole. This does not mean, as has sometimes been claimed, that there is a fundamental shift in Heidegger's later work which reflects a profound change in his understanding of temporality. Instead, his relieving temporality of any existential significance should be regarded as a framework within which Being can be *given* in various ways. In other words, time becomes that which gives or makes possible Being, that from which Being arises.

There are several important parallels between Heidegger's and Derrida's understanding of time, three of which will be highlighted here. First, they share a general analysis of the history of philosophy as a history of the dominance of *presence*. For both men, presence is constituted by something that is not presence itself. Whether or not this "something" is called originary temporality, it appears to have temporal qualities, and it clearly undermines the dominance of the vulgar concept of time as constituted by a succession of nows. Second, both philosophers claim, in a way that differs entirely from the relativity theories of physics, that temporality and spatiality are not fundamentally separate concepts. Derrida connects them with *différance*, Heidegger with his concept of space-time, which he set out in the essay "Time and Being" (1962).[8] Third, and of special importance for Derrida's own thinking about time, from the mid-1930s Heidegger tried to differentiate between originary temporality and human existence or *Dasein*, so that time might instead be understood as *event*. We might say that Heidegger moves the centre of gravity, from a description of time as structure to an engagement with time as event. This engagement means that Being is no longer subordinated to time; rather, Being and time represent each other's condition of possibility. According to Derrida, the German term for event, *Ereignis*, which is difficult to translate, is strongly associated with the French word *appropriation* (appropriation), which in turn is closely tied to the concept of *don* (gift).[9] In this way, Being becomes directly connected to the gift. This movement, which is already indicated in *Being and Time*, takes on a special significance in "Time and Being" by means of Heidegger's analysis of the German expression *es gibt*. Heidegger uses the expression in the sense of *es gibt Sein* and *es gibt Zeit*,

[8] Wood, *The Deconstruction of Time*, pp. 259, 263

[9] Derrida, *Donner le temps 1. La fausse monnaie*, pp. 33–4. The word *Ereignis* (event) appears as early as Heidegger's first lectures in 1919 (see *Gesamtausgabe* vol. 56/57). It recedes in *Being and Time*, only to re-emerge as the central theme of *Beiträge zur Philosophie* (*Vom Ereignis*), written in 1936–1938 (see *Gesamtausgabe* vol. 65). Heidegger's philosophy thereafter becomes a thinking through of *Ereignis*, and the words has a prominent place in several of postwar works, particularly the late essay "Time and Being".

which might be translated as "Being is given" (Being *is* not, but it is given as being) and "time is given" (time *is* not, but time *is given*). It is this very giving, this relation between time and the gift, that Derrida develops further, above all in his central work on time, *Donner le temps 1: La fausse monnaie* (1991).

Heidegger and Derrida differ on one fundamental issue, however. For Derrida, the central idea in *Being and Time*, namely, contrasting a non-metaphysical originary time with a supposedly "vulgar" time, is problematic to say the least. He summarizes his analysis in *Ousia et Grammè* by wondering whether a "vulgar conception of time" actually exists. Is it really possible to contrast *another* concept of time to the "vulgar" one, when time, as he sees it, generally belongs to metaphysics? Why does Heidegger propose a kind of *fall* from an *originary* to a *derivative* temporality? Is this longing for origins, for a firm basis, not the central feature of all metaphysics? Is there not a kind of Platonism in this *Verfallen* (lapse)? Derrida's conclusion is that there is not (and cannot be) a non-metaphysical, originary temporality. The only time that exists is the one that Heidegger calls "vulgar." And this time, as we have just seen, is in its very essence *aporetic*. In *Aporias* (1992), which can be seen as a continuation of Derrida's reading of Heidegger in *Ousia et Grammè*, he goes a step further and combines the aporia *as such* with time:

> What if the exoteric aporia therefore remained in a certain way irreducible, calling for an endurance, or shall we rather say an *experience* other than that consisting in opposing, from both sides of an indivisible line, an other concept, a nonvulgar concept, to the so-called vulgar concept"?[10]

By connecting *experience* with the concept of *permanence*, Derrida gives experience an explicitly temporal character. His intention is clearly to formulate a line of thinking about the experience of the *impossible*. Derrida wonders: what would such experience involve? Is it possible to speak of—and, if so, in what sense—an *experience of aporia*? An experience *of aporia as such*? Or, conversely: "Is an experience possible that would not be an experience of the aporia?"[11]

4.4 Derrida's Deconstruction of Time

In *Psyché: Inventions de l'autre* (1987), Derrida explicitly defines the aporetic experience of the impossible as *deconstruction*.[12] Because this experience is *permanent*, deconstruction is essentially aporetic thinking *in time*. Deconstruction is continually engaged with time, particularly the future, at the same time as it is transforming the concept of time. This transformation is realized by the inscription of time in space, its *becoming-space*.

---

[10] Derrida, *Aporias*, p. 14

[11] Ibid., pp. 14–15

[12] Derrida, *Psyché: Inventions de l'autre*, p. 27.

But if *becoming-space* means that time is inscribed as spatially separated nows, how then are we to connect the past to the future? How, in other words, is identity achieved across time? Philosophers of consciousness typically resolve the problem with the help of a *synthesis* of past and future nows produced by a self-present subject, which is to say a subject who is present to itself in the immediacy of the psychic act. This immediacy, this idea of the subject's self-presence, is what Derrida calls into question. For Derrida, the now is not an immediate now and its constitution is not *simple*. On the contrary, the now is an originary and irreducible synthesis that is *not* grounded upon the subject's self-presence. This synthesis presupposes that the now is *divisible*, allowing the creation of an "interval" that separates the now from what it is not. It is at this point that Derrida introduces the concept of "trace" as that which constitutes the now in an originary synthesis:

> In constituting itself, in dividing itself dynamically, this interval is what might be called *spacing*, the becoming space of time or the becoming time of space (*temporization*). And it is this constitution of the present, as an "originary" and irreducibly nonsimple (and therefore, *stricto sensu* nonoriginary) synthesis of marks or traces ... that I propose to call archi-writing, archi-trace, or *différance*.[13]

Derrida here connects the trace to *espacement*, that is to say, to the becoming-space of time and the becoming-time of space, which is also the definition of arche-writing and *différance*. This definition thereby makes possible an entirely new conception of temporality, here understood as an originary synthesis of inscribed traces or signs and thus *not* as a synthesis produced by a self-present subject.

What exactly does Derrida mean by the *becoming-space of time*? Because the current now perishes at the very moment it comes into existence, it must be split up and inscribed as a trace in space simply in order to endure. This is the *becoming-space of time*. The now must be inscribed as a trace in space because spatiality is characterized by the capacity to endure despite the fleeing character of time. Spatiality is thus a necessary condition for the synthesis because it makes it possible to connect the past to the future. Yet spatiality cannot endure by itself; the inscribed traces cannot be simultaneous. Simultaneity can only exist *as such* by means of a temporality-creating synthesis that connects one spatial position to another. The *becoming-time* of space is necessary, not only in order to connect the trace to other traces but in order for traces to even exist. A trace can only be read after it has been inscribed, whereupon it is permanently defined by the *relation* to the future that temporalizes space.[14]

If time must be inscribed spatially, the consequence is that *time must be related to the technological*. Indeed, as Martin Hägglund has recently argued, the experience of time is essentially dependent upon the material support and technologies that are available for the inscription of time.[15] In other words, there is no "pure" time, separate from the technological: time is always mediated by the

[13] Derrida, "Différance", pp. 13–14.

[14] Derrida, *Ousia et Grammè*, p. 63.

[15] Hägglund, *Radical Atheism*, p. 27.

technological-material. For a temporal event to exist, *something* must happen, some-*thing* must inscribe traces in space. This inscription does not take place in an already constituted space; rather, the inscription produces the very spatiality of the space. Derrida is thereby able to claim that time and space are not "pure" transcendental modes of thinking, as in Kant, but are constituted through technological mediation. Because *espacement*—like arche-writing and *différance*—are fundamental conditions of possibility, the inscription of time must *always already* be mediated via the technological-material. Technological inscription can be changed, manipulated, and erased in all kinds of ways, but *espacement* as a fundamental condition of possibility remains.

This fundamental condition of possibility, which we will consider in greater depth in the next chapter, is in no way limited to human beings. In *De la grammatologie*, Derrida explicitly defines *différance* (and thereby *espacement*) as the history of life in general, in which the history of life is always conditioned by different forms of inscription, ranging from the "'genetic inscription' and the 'short programmatic chains' regulating the behavior of the amoeba or the annelid up to the passage beyond alphabetic writing to the orders of the logos and of a certain *homo sapiens*."[16] He emphasizes, in other words, the fundamental character of inscription in its most general sense as comprehending all forms of genetic and technological inscription.

It is through inscription and synthesis, through the movement of *différance*, that the traces of the past are connected with the traces of the future, with that which is to come, with that which will come to pass. For Derrida, this future is always uncertain and unforeseeable even as it comes unexpectedly and inconveniently and therefore demands prompt decision-making. Thinking about *différance* is therefore also thinking about haste, about the fact that I cannot escape from the unique thing *that is happening* here and now, that is to say, the individual *event*. What is an event? It is, says Derrida, simply an experience. Not any old experience, but *experience itself*. The event is another name for what we can neither reduce or deny (or flatly deny) in that which is happening. It is another name for experience itself, which is always an experience of the other. The event cannot be subsumed under any other concept, not even under Being. The fact that "there is" ["*il y a*"] or "there is something rather than nothing" are perhaps part of the experience of the event rather than a way of thinking about Being.[17]

For Derrida, the arrival (*le venir*) of the event, which may not be obstructed, no matter what, is not only an arrival from the future, it *is* the future itself. Not the future in the usual temporal sense, as a future now, but that which comes to us from without, which is always already in the future.[18] This absolute or structural future—Derrida uses the French word for the future, *avenir*—constitutes a powerful tool for changes in the present. Derrida often refers to justice to come in the form of a

[16] Derrida, *De la grammatologie,* pp. 125–6.

[17] Derrida & Stiegler, *Échographies de la télévision*, p. 19.

[18] According to Derrida, the future can "always reproduce the present. Justice remains, is yet, to come, *à-venir*, it has an, it is *à-venir*, the very dimension of events irreducibly to come. It will always have it, this *à-venir*, and always has". (*Force of Law*, p. 27).

promise, and, as with a promise of a coming Messiah, it is a tool by which the future can intervene and affect the present. At stake here is our relation to that which is entirely other, what Derrida, borrowing a term from Benjamin and Levinas, calls the *messianic*; the messianic can come at any time, suddenly, like a bolt from the blue. Such a situation demands an openness to the arrival of that which is entirely other. It is therefore as a way of thinking about time that the event "always opens a certain messianic space," which is why "this messianic dimension cannot be separated from justice."[19]

Time as event opens up a messianic space in which the word "come" (*viens*) can be spoken to the other or others. The event presupposes that this "come" is spoken to someone, to "someone else that I cannot and must not determine in advance, not as subject, self, consciousness, nor even as animal, god, or person, man or woman, living or non-living thing."[20] The event is unpredictable and cannot be reduced to the fact that something is happening. Derrida takes the weather as an example: "[I]t may rain tonight, it may not rain. This will not be an absolute event because I know what rain is."[21] An absolute event is instead something that suddenly comes, something that unexpectedly occurs, something that by definition I know nothing about in advance. Derrida's understanding of time as event therefore involves nothing less than a revolution in the classical concept of time. The event represents an absolute rupture in the predictable homogeneity of historical time, in which messianic eschatology intervenes and interrupts teleology.[22]

One of Derrida's many names for this "something" that suddenly comes and interrupts the homogeneity of time is *the gift*. As mentioned earlier, he connects the gift to time in *Donner le temps*. And he does it, as he himself says, for *economic* reasons. In his view, both the gift and time are directly related to the economic. He begins by explaining that economics is about exchange, circulation, return. The figure of the *circle* is undeniably a key feature of the economic: we talk about circulation of goods and services, financial transactions, mortgage repayments, incomes, and so forth. This motif of circulation is central to an understanding of what economics fundamentally is: a circular return to the starting point, to the point of origin, to the same.[23]

The gift is unquestionably related to the economy, even the money economy. But a *pure* gift, if such a thing exists, is hardly given in order to be returned. It is not part of a closed economic circuit but, on the contrary, is precisely that which *interrupts* the economic, that which by suspending the economic calculus no longer leads to exchange. The gift breaks the circle in order to challenge reciprocity and prevent a return to whatever it is that gives. If economics is characterized by the figure of the

[19] Derrida & Stiegler, *Échographies de la télévision*, p. 19.

[20] Ibid., p. 20.

[21] Ibid., p. 21.

[22] Ibid.

[23] Derrida, *Donner le temps*, p. 18.

circle, the pure gift must be the opposite of the economy, it must be *aneconomic* (*anéconomique*).[24]

As noted earlier, time has also been represented by a circle or a sphere: "One of the most powerful and ineluctable representations, at least in the history of metaphysics, is the representation of time as a circle. Time would always be a process or a movement in the form of the circle or the sphere."[25] It means that time, or what Heidegger calls "vulgar" time, is absolutely incompatible with the gift:

> Wherever time as circle … is predominant, the gift is impossible. A gift could be possible, there could be a gift only at the instant an effraction in the circle will have taken place, at the instant all circulation will have been interrupted and on the condition of this instant. … This condition concerns time but does not belong to it, does not pertain to it without being, for all that, more logical than chronological. There would be a gift only at the instant when the paradoxical instant (in the sense in which Kierkegaard says of the paradoxical instant of decision that it is madness) tears time apart. In this sense one would never have the time of a gift. In any case, time, the "present" of the gift, is no longer thinkable as a now, that is, as a present bound up in the temporal synthesis.[26]

A gift that is given in chronological time is therefore an impossibility. If the gift is an event in which someone gives some-*thing* to someone else, the possibility of the gift must also be its impossibility. In other words, if the gift is defined by the formula A gives B to C, it negates, annuls, and destroys itself *a priori*.[27] If the gift is absolutely pure, then it has nothing to do with reciprocity, repayment, reimbursement, or debt. If the other *gives back* in kind or *owes* me or must give back what I give, then no gift has occurred, regardless of when repayment takes place. If someone gives back the same thing—for example, an offer of lunch—the gift ceases to be. The gift that comes from the future, by contrast, is not determinate or something that has a particular given value. Nor is it something that should necessarily be regarded as *good*. On the contrary, it may reveal itself to be unwelcome and can even be a bane (a word that also means poison, as does the German word for gift, *Gift*) from the very moment that the gift places the other in debt, which is to say, causes the other injury.[28]

As Derrida himself stresses, his definition of the gift differs from that of anthropologists and philosophers, who traditionally regard the gift and the return gift *together* as a closed system. By contrast, he argues that there is gift, only by virtue of the fact that it disrupts the system and all that it represents.[29] The gift thereby also disrupts the temporality of time with its memories, expectations, retentions, and protentions. If the gift can be considered an event, it must take place in a time without time, an un-time, a "now" without now.

---

[24] Ibid., p. 19.

[25] Ibid.

[26] Ibid., p. 21.

[27] Ibid., p. 23.

[28] Ibid., p. 25.

[29] Ibid., pp. 25–6.

This line of thinking with regard to the time of the gift can be understood as an attempt on Derrida's part to distance himself from Husserl's analysis of the givenness of the phenomenon and to instead embrace an analysis of this paradoxical "now" that is not a now, of a givenness that cannot be predicted and that cannot be repaid without nullifying the status of the given *as gift*. This displacement, from the given to the gift, can be seen as Derrida's attempt to extricate himself from what interests Husserl, namely givenness as a foundation that guarantees truth and certain knowledge. While Derrida distances himself from Husserlian discourse, he nonetheless builds, as we have already remarked, upon Heidegger's meditations in *Being and Time* and later works upon the expression "*es gibt*" (it is given, there is), as in the phrase "it is given time." For Derrida, the gift is precisely that which *gives time*. The difference between gift and return gift is that the gift is that which gives time. But the gift must also be allowed to *take* time. In other words, the time between gift and return gift must be of a certain duration—a certain temporal *différance*—that cannot be either instantaneous or infinite. Rather, it must be determined by its own rhythm, a rhythm that has not been derived from homogeneous time but that structures time at its origin. The gift is an event that opens up the closed homogeneous system of time to another time that is more complex and qualitatively more heterogeneous. And such an event is *artificially produced.*

4.5 The Artifactual Production of the Event

In 1996, Derrida published a book that contains perhaps his most detailed examination of the media and, in particular, television: *Échographies de la télévision.*[30] In my view, this book should be regarded as showing deconstruction's engagement with how media, with help of technology, *produces time*. Right from the introduction, Derrida affirms that time "today, more than ever before," is calculated, constrained, "formatted," and "initialized" by a *media apparatus*. Time is, in fact, produced *artificially*: "It is an *artifact*."[31] It is being continually created by a gigantic "teletechnological apparatus" that includes not only media in the form of the press, radio, TV, and film but also mobile communication systems, digital archiving

[30] Derrida was a frequent guest in television studios around the world, but he seems also to have been an inveterate television viewer: "Aside from the news I watch all kinds of things, the best and the worst. Sometimes I watch bad soap operas, French or American, or programs that give me a greater cultural awareness ... political debates, spectacular political encounters in general ... or else good movies. I could spend 20 h a day watching good political archives ... And so I watch a little of everything.... I watch very regularly Sunday morning ... the Muslim and Jewish religious programs, which I find very interesting....Then comes the Christian hour: Orthodox, Protestant, then Catholic." (*Échographies de la télévision*, p. 155).

[31] Derrida & Stiegler, *Échographies de la télévision,* p. 11.

systems, and a global infrastructure in the form of industries, experts, treaties, and work patterns.[32]

This teletechnological apparatus creates the event, that which happens, "actuality." Derrida uses the French word *actualité*, which could be translated as "now" or "the present" were it not that he precisely differentiates between *actualité* and the originary now that disappeared at the very instant when it appeared.[33] Engaging with the now or the present, for example, as a political philosopher, therefore seems to presuppose a distinction between actuality and the originary now. Derrida identifies two attributes of actuality, which he designates using two neologisms: *actuvirtuality* and *artifactuality*. Actuavirtuality emphasizes the *virtuality* of the actuality. Derrida says that he "insists" upon this virtuality.[34] The virtual image, the virtuality of time and space, the virtual event—basically, everything designated by the English term *virtual reality* can no longer be contrasted with a supposed "real" reality:

> [The] differantial deployment of tekhne, of techno-science or tele-technology … obliges us more than ever to think the virtualization of space and time, the possibility of virtual events whose movement and speed prohibit us more than ever … from opposing presence to its representation, "real time" to "deferred time,"[35]

This form of virtuality, which Derrida also calls "teletechnological *différance*,"[36] inevitably defines the very structure of the event that is produced, affecting time and space in everything that we associate with an underlying "real" reality. In other words, we must be sensitive to the consequences of this virtual space-time. Not only does Derrida refer to this actuvirtuality, he also points out the fact that actuality is in fact *artifactual* in that it is artificial or *made*:

> In order to know what it [actuality] is made of one needs nonetheless to know that it is *made*: … It is not given but actively produced, sifted, invested, performatively interpreted by numerous apparatuses which are *factitious* or *artificial*, hierarchizing and selective, always in the service of forces and interests to which "subjects" and agents … are never sensitive enough.[37]

It is necessary to learn to understand how newspapers and television news programmes are *made* and *by whom*. For instance, when the television newsreader speaks directly to us at home, looking us straight in the eyes, he or she is in fact reading a text on a screen, a text produced *at another moment in time*, somewhere else, perhaps by an entire network of anonymous producers.

According to Derrida, the deconstructive analysis of the conditions for actuality's technological and artificial production—"the requisite deconstruction of this

---

[32] Today, this "teletechnological apparatus" would also include social media such as Facebook, You Tube, Twitter, etc.

[33] Its plural form *actualités* means simply "news". In Sweden, there is even a current affairs programme called *Aktuellt*.

[34] Ibid., p. 14.

[35] Derrida, *Specters of Marx*, p. 212.

[36] Ibid., p. 238, n. 12.

[37] Derrida & Stiegler, *Échographies de la télévision*, p. 11.

artifactuality"[38]—must be attentive to at least three problems.[39] First, the problem of the *national*. Despite intensifying internationalization, national, regional, and local news predominate in the selection process that characterizes actuality (*actualité*). In news programmes, actuality is spontaneously ethnocentric. There is a tendency of excluding the alien, the other. Second, and in contrast to the problem of the national, the accelerating internationalization of information sources is often based on a growing concentration of information and communication capital. When that happens, as is currently the case, a handful of global television players come to dominate the modern TV wars. International artifactuality, the monopolization of the "actuality effect," and the concentration of the artifactual ability to create an event are often followed by *live* broadcasts. For Derrida, it is the task of philosophy to show that what is supposedly reaching us *live* is never absolutely *live*, but rather a *live* effect that never gives us the originary now freed of all interpretation and technological manipulation. It is important to understand that what is broadcast *live* on a TV channel is "*produced before being transmitted.*"[40]

Third, the deconstruction of artifactuality must not become an alibi for "the delusion of the delusion," a denial of the event. Showing how reality and the event are produced and presented in purely technological terms, and showing that there are no facts—that is, events whose meaning is once and for all determined, archived, and fixed—do not legitimate fictionalization: analysing how the event is fettered to technology does not mean dissolving it into pure fiction. The fact is that technology produces the reality that it presents (and that Derrida calls artifactuality), but it never produces the originary event, which can only be certified by whoever was a witness, by testimony. A proof that is generated technologically is not testimony and can never serve as proof of an event, something only testimony can do: "Technics will never produce a testimony."[41]

In other words, deconstructive analysis is deeper than the somewhat strange interpretation of reality proposed by Jean Baudrillard. For instance, Baudrillard seems to be arguing that the Gulf War (1990–1991) did not take place, that it was, for us, *only* medial or virtual:

> What Baudrillard meant …[was] that, precisely, the simulacra of images, television, the manipulation of information, reportage had nullified the event, that in the end this was lived only through the simulacrum. This is interesting. I believe something like this or something analogous happened (and no doubt always happens, has always happened, from the moment that iterability in general structures the eventness of the event).[42]

Deconstruction therefore finds that the simulacra of the event are inscribed into the very centre of the event structure: the event itself becomes a simulacrum, it is "nullified" from the moment it becomes an event (by repeating itself informationally

[38] Ibid.

[39] Ibid., pp. 12–14.

[40] Ibid., p. 49.

[41] Ibid., p. 107.

[42] Ibid., p. 89.

and by thereby distancing itself from itself). It is, in fact, production and repetition that create the event: in the precise moment when the event repeats itself, it is changed and ceases to be unambiguous or identical with itself. Derrida calls this repetition "iterability." This changing of the event through repetition and iterability arises as soon as archiving exists.

The event is thus nullified by means of its own movement: it retreats because it can only appear as a result of repetition and change. The event is actually only "event" if it is interpreted and understood as a repetition that confers the status of an event. This structural withdrawal requires that deconstruction take responsibility for the memory of that which has actually taken place; it is because the event has been erased from memory that we must work not to forget it and must never forget the erasure of that constitutive memory. What Baudrillard says

> should not make us forget—and the event is unforgettable—that there were deaths, hundreds of thousands of deaths, on one side of the front and not the other, and that this war took place. … [W]e should not forget that these deaths are each time, by the hundreds of thousands, singular deaths. Each time, there is a singularity to murder. It happens, and no process, no logic of the simulacrum can make us forget this. For, along with process, we must also think singularity.[43]

The deconstruction of actuality and its production is therefore *aporetic*. It cannot be satisfied with the virtual event but must also conceive of the impossible: the originary event as it actually took place. But aporia cannot become dialectical: the rigid opposition between the virtuality of the event and the origins of the event cannot be sublated into a kind of Hegelian synthesis. The way that deconstruction takes responsibility for the event is by persisting in a continual reflection upon this contradiction. Taking responsibility like this, which involves simultaneously analysing the originary and the virtual event, is only possible through active archival research. For this reason Derrida is able to write: "The fact of having access to these archives, of being able to analyse their content and the modalities of selection, interpretation, manipulation that superintended their production and circulation, all these rights are therefore a citizen's right."[44] Active participation and critical scrutiny of politics, which are preconditions for real democracy, are not possible without this unconditional and universal right to access the archive.

The deconstruction of the archive presupposes technological knowledge of and taking responsibility for the event; it is, Derrida argues, a civil right rather than a philosophical practice or an abstract theory. In fact, deconstruction *acts*, particularly when it is working to analyse the event and trying to understand how the event has been produced or archived. It is really an issue of democracy: the civil right to deconstruct the archive and the archived artifactuality is a democratic right that cannot be separated from the democratization process. According to Derrida, there can be no democracy without this possibility, and the right to deconstruction is not a

---

[43] Ibid.

[44] Ibid., p. 45. Derrida is here referring to the right and the duty of society, the state, or the nation of storing and preserving essentially everything that is produced and broadcast via the national media.

philosophical right but the name for democracy's actual movement. It is therefore "completely normal for the state to guarantee, without delay, to anyone, who wants to study these public documents—one thinks first of researchers and of research that ought to be developed in the audiovisual domain—not only the formal right, but the technical conditions of access to this archive. If the enforcement of this law has been delayed, this is unacceptable. … [T]he delay in this domain is a violation of the right of anyone who wants to consult a public archive."[45]

Technology is here understood as the condition for access to the archive and, through it, the event, precisely because the event cannot happen without technological reproducibility: the event's technological reproducibility in media and archives, its "artifactuality," constitutes its "eventness." There is no event that has not been reproduced, because the originary event was immediately eradicated. The dramatic development of archiving and communications technology therefore involves a democratization of access to the archive, because this new technology constitutes the practical deconstruction of a certain political power. This, Derrida argues, will speed up the current process of global democratization: "I believe that this technical transformation—of the telephone, of the fax machine, of television, e-mail and the Internet—will have done more for what is called "democratization," even in countries in the East, than all of the discourses on behalf of human rights, more than all the presentations of models in whose name this democratization was able to get started."[46]

For Derrida, then, technology is a tool for a powerful practical deconstruction of time. Its democratic ramifications cannot be exaggerated and may prove to be considerably more far-reaching that what has been achieved by all the eulogies to democracy and human rights. This fact, he hopes, may accelerate the global process of democratization and pose a threat to all forms of totalitarianism.[47] Beyond all moral judgements about technology, Derrida tries to understand the transformation of the political *through* technology; his analysis of technology can be described as realistic in the sense that he does not condemn technology or call for a halt to technological development. In fact, he does not present the non-technical as the opposite of the technological. Nor is deconstruction opposed to technology as such; rather, it distinguishes between different technologies' potential to promote or retard the democratization process. If technology thus cannot be represented as the opposite of something non-technological that supposedly precedes technology, it is because "preceding" is already technological: "These machines have always been there,

---

[45] Ibid., p. 44.

[46] Ibid., p. 82.

[47] The democratization process must be seen as a very long-term, not to mention never-ending, process. It took almost two hundred years for democracy to develop in the West—from the ideas of the Enlightenment to the adoption of universal suffrage in the years after the First World War. Applied to international politics, this perspective implies that it will take centuries before a comparable democratic system is realized at a global level. Like many other commentators at around the time of the fall of the Berlin Wall, Derrida here seems a little too optimistic about technology's potential to accelerate the democratization process. (My notes.)

they are always there, even when we wrote by hand, even during so-called live conversation."[48] Technology is quite simply that which stirs up time, the time that shows what the hour is. "The time is out of joint," declares Hamlet in his famous monologue, a line that serves as the central motif of *Specters of Marx* (1994). For Derrida, too, time is "out of joint," disjointed or disadjusted, *disajointed*.[49] Reflecting upon this *anachronistic* time also involves reflecting upon technology, upon the disturbance of chronology through technology.

4.6 Technology Puts Time out of Joint

There is a widespread view that time is speeding up in modern society, indeed, for some, that it is even going too quickly.[50] No one doubts that this change in our rhythms of living is closely bound up with technology and technological development. It is, then, the acceleration of technology and acceleration through technology itself that are making everything, objectively, go so very much more quickly.

But technology is not only that which makes things go more quickly; it is not in the first instance something that radically reduces time. Nor does it dissolve distances in space or chronological time as such. Technology is, above all, that which *disturbs our experience* of time and space. This disturbance of time through technology, which literally puts the future before the past and the past after the future, causes what Derrida calls *the spectre* to show itself.[51] The spectre or phantom shows itself as soon as technological inscription or archiving exists, for example, in the form of a photographic image: "As we know, once it has been taken, captured, this image will be reproducible in our absence, because we know this already, we are already haunted by this future, which brings our death. Our disappearance is already there."[52] This notion of the ghost or phantom as *produced* by the technological apparatus is one of the most surprising and original ideas in Derrida's deconstructive thought: it is undoubtedly a result of his thoughts on technology. But why does technology produce spectres and phantoms?

---

[48] Ibid., p. 47.

[49] Derrida, *Specters of Marx*, p. 1. A few pages on, Derrida adds that "time is disarticulated, dislocated, dislodged, time is run down, on the run and run down [*traqué et détraqué*], deranged, both out of order and mad. Time is off its hinges, time is off course, beside itself, disadjusted" (p. 20).

[50] The German sociologist Hartmus Rosa even considers speed and acceleration to be the very kernel of modernity, the central concepts of our time. (See his *Social Acceleration: A New Theory of Modernity*, 2013). A similar argument has been made by the French architect, urbanist, and philosopher Paul Virilio.

[51] Derrida and Marx were not the only ones to have spectres. Even Freud and even Heidegger had theirs, it seems: "The most cultivated, the most reasonable, the most nonbelieving people easily reconcile a certain spiritualism with reason … Freud had his ghosts … and he obeyed them … as do I myself (Derrida, *Mal d'archive*, p. 138–40). On Heidegger's "spectres", see Derrida & Stiegler, *Échographies de la télévision*, p. 147.

[52] Derrida & Stiegler, *Échographies de la télévision*, p. 131.

It comes from the fact that deconstruction, as noted earlier, is fundamentally an aporetic mode of thinking. It does not approach reality simply as actual reality, in contrast to virtual, but inscribes the possibility of something entirely other, "of radical alterity and heterogeneity, of *différance*, of technicity, and of ideality in the very event of presence."[53] Technological inscription, what Derrida acutely calls "the prosthetic synthesis,"[54] thus enables deconstruction to observe the effects of the virtual in the form of spectres and phantoms.

What is a spectre? The spectre is not only that which observes us but that we cannot observe, it is not only what Derrida in *Specters of Marx* calls a *visor effect*,[55] that which elicits a feeling that someone is watching us, that someone else is taking an interest in us. Spectres are also revenants (*revenants*) that appear out of the past: they shake up chronology because in them the past becomes the future. This apparition should nonetheless not be imagined as a liberation of the idea or the thought from the bodily, as happens in Hegelian idealism. Rather, it is that when ideas or thoughts have been liberated "one engenders some ghost by giving them a body. Not by returning to the living body from which ideas and thoughts have been torn loose, but by incarnating the latter in *another artifactual body, a prosthetic body.*" This prosthetic body, which is both visible and invisible, is not flesh and blood but "a technical body or an institutional body."[56]

These phantoms and revenants—of which, according to Derrida, there are more than one—appear, as we have seen, from the moment that technological inscription and archiving come into existence. We have also seen that what is inscribed, from this moment, can also be reproduced and repeated indefinitely. Reproduction and repetition act through writing, photography, film, radio, and television but also in speech, hearing, and memory: in some sense, everything that is produced technologically can, from the moment of its production, also be reproduced. Because these traces can be reproduced in the absent of the producer, technological reproducibility is a fundamental potential of the trace, which is why the trace must be regarded as an revenant. Technical reproducibility haunts us: it causes us to have a presentiment about and, perhaps, to experience—as in the photographic image—the possibility of our own demise, our own absence.

Technical reproducibility is a revenant in the strict sense of the word. It makes spectres appear and it makes the past appear in the now as that which is to come. If I look at a photograph, I realize that the image will outlive me, which it in fact already does from the moment it "is taken." What I am seeing in the photograph is me, but from another perspective it is no longer me but someone else; it is me as I will be thought of after my death, because the image of me is already outliving me from the very moment when it is created (if it is not destroyed). The inscription of my demise in the photograph and the continued existence of this image is about

[53] Derrida, *Specters of Marx*, p. 94.

[54] Ibid.

[55] Ibid., p. 121.

[56] Ibid., pp. 157–158.

"me" yet not *me*; the image observes me, it haunts me, but I cannot take it for myself. In other words, my entire "living now" has been shaken up by the spectral appearance of this technological reproducibility. The living now is preceded by this pre-posthumous spectrality that acts through writing, speech, photography, film, etc.

In *Ghost Dance*, the film in which Derrida appears as himself, he says: "Modern technology, contrary to appearances, although it is scientific, increases tenfold the power of ghosts."[57] The revenants come back but in another guise, changed, different. They are the event that returns but in a changed form; through the possibility of repetition, it returns in a different context to that which is given, it transcends, in other words, all *given* contexts.

The phantoms that technology calls forth move through time, they are marked by infinity; these phantoms are not outside of time or eternal, they are finite. In a television recording with Bernard Stiegler, Derrida meditates upon the finiteness of life: "No matter how soon after their recording we die, and even if we were to die while recording, *voilà*, this will be and will remain 'live,' a simulacrum of life. A maximum of life (the most life [*le plus de vie*]), but of life that already yields to death ('no more life' ['*plus de vie*']) , this is what becomes exportable for the longest possible time and across the longest possible distance—but in a finite way. It is not inscribed for eternity, for it is finite, and not just because the subjects are finite, but because the archive we're talking about, too, can be destroyed."[58] In the midst of analysing technology and phantoms, that is to say, analysing things that can seem unchanging or dead, Derrida's texts are engaged in thinking about the continuation of life, albeit in other forms.

Derrida thus sees technology and the spectres that it conjures up, not as a dead mechanical mechanism but as a *plus de vie*, a sur-vival (*sur-vie*). Technology confers on life a *plus de vie*, a supplement to life, a life that returns in other ways. Technology is in this sense "living." Derrida's understanding of it is anything but mechanistic; on the contrary, technology allows life to remain living beyond life, through another kind of life, a spectral life. In this sense, technological processes are finite, here understood as destructible: we must imagine, according to Derrida, a total destruction of the archive that leaves no trace of its traces. By producing another life, and by inscribing in it a *plus de vie*, technology also confers upon life a finiteness other than that of the living: their spectral finiteness, their potential disappearance and eradication.

57 Derrida & Stiegler, *Échographies de la télévision, op. cit.*, p. 129.

58 Ibid., p. 48.

4.7 Stiegler Versus Derrida: Technological or Messianic Time?

In modern philosophy the question of technology's relation to time is associated with one name in particular: Bernard Stiegler. His fame in the world of philosophy of technology stems primarily from his three-volume magnum opus *La technique et le temps* (1994, 1996, 2001). There are therefore good reasons for making a critical comparison of Derrida's and Stiegler's respective positions on this question.

Bernard Stiegler is very much the heir of his mentor Jacques Derrida. He takes his point of departure in Derrida's account of the grammatology of arche-writing, which he develops into a general philosophy of technology in which technology plays a crucial role in human life. For Stiegler, human beings should be regarded as *Homo faber* and technology as their gradual *externalization*. Technology is instrumental and encompasses everything from primitive tools to various writing systems and modern telecommunication. Thus far Derrida and Stiegler are in agreement.[59] But technology is not only instrumental, it also determines what it is to be human; the human is always already based on technology. Stiegler argues that human beings are created *by* technology even as they create it. Human beings and technology are essentially two elements in the same dynamic feedback system.

Stiegler views technology as a prosthesis: unlike animals, human beings exist in a prosthetic relation to technology, to what Stiegler calls "organized inorganic matter." According to Stiegler, this prosthetic relation involves a radical interruption in genetic evolution and is the start of human development, which is a development towards increasingly technologically advanced means of assistance. Technology is, for Stiegler, "the pursuit of life by means other than life."[60]

Perhaps the most original aspect of this positivistic philosophy of technology is its account of "tertiary memory." Stiegler argues that Husserl's thinking excludes externalization of human memory in the form of mnemo-technological aids. Although these aids are empirically derived from human memory, Stiegler argues that from the beginning they have constituted the way in which we think about the past, the present, and the future. We humans are constituted *a priori* by these technological aides-mémoires. Stiegler calls this form of memory "tertiary memory." It comes after humans' primary and secondary memories in *empirical-historical* terms but precedes them *structurally*. Regardless of the historical epoch, therefore, technological-material aids produce a predetermined inscription in human memory from which we temporalize ourselves and our world. Stiegler thereby inscribes modern phenomenology within a *materialist* history even as he abandons *the trace* and Derrida's entire quasi-transcendental logic. This innovation undeniably represents a form of technological determinism that regards technology as constitutive of

[59] Derrida and Stiegler are naturally also in agreement about the problematical nature of Heidegger's statement that "the essence of technology is nothing technological".

[60] Stiegler, *Technics and Time 1*, p.17.

humanity's conditions of possibility while also reducing those conditions to technological aids.

In keeping with this materialist philosophy of technology, Stiegler argues that the human capacity to temporalize—to organize experiences in time—is constituted, rather than mediated, by mnemo-technological prostheses. Quite simply, our experience of time has its origin in technology. Stiegler is thereby able to formulate his main thesis in *La technique et le temps*: *all* time is entirely technology.[61] It is not enough, therefore, to try to understand technology *in* time, that is to say, technology's own historical development, nor to see it as a question *of* time, of our experience of time, but as something even more radical: we need to understand technology *as* time: technological aids are what make possible our temporalization of time in the past, the present, and the future.[62]

There is clearly an unresolvable conflict between Derrida and his student with regard to the relation between technology and time, a conflict that appears most acutely in their co-authored book *Échographies de la télévision*. Throughout this book Derrida argues that *différance* cannot be reduced to technology. The technological apparatus does not represent, as Stiegler seems to argue, a break with the quasi-transcendental logic of *différance*. On the contrary, this apparatus is a necessary supplement to the human. Technology is supplementary for Derrida, constitutive for Stiegler.

Their differing views of technology's status mean that Stiegler has a different philosophical agenda to Derrida. Because Stiegler regards technology as originary and constitutive, he tries to understand technology by describing its *empirical* history, from flint tools to modern digital technologies. Derrida, on the other hand, develops an entirely new *quasi-transcendental* logic, a logic based on *différance*, according to which technology is *a priori* a necessary supplement to the human.

As we saw in the previous chapter, Derrida argues that it is problematic for Stiegler to claim that we can understand technology if that technology is itself "originary" and constitutive. It is quite simply both logical and topologically impossible to reduce the quasi-transcendental level to technology.

Derrida's way of developing the question of the relation between technology and time differs from Stiegler's in another respect, too. Derrida counterposes an uncertain and unforeseeable messianic time to a homogeneous, historical, or chronological time. For Stiegler, any attempt to view these two concepts of time as opposites is outright Platonism: all time is technological time. In our current capitalist society, as Stiegler sees it, the future is technologically determined. He necessarily views Derrida's talk of humanity's openness to the future, to that which is entirely other—"the messianic"—as problematic: in our hyper-industrialized era, human beings are de-individualized, "is no longer anything but purchasing power, that is blind consumption." The individual has been turned into a proletarian, "has become a serf

[61] Ibid., p. 27.

[62] Ibid., p. 83.

again."[63] There seems to be a yawning chasm between Stiegler's form of technological determinism and Derrida's openness to the future. On the one hand, if humanity's future is determined by the teletechnological apparatus, should it not be politically managed in the interests of humanity? On the other hand, if this apparatus is a supplement to humanity, would a new "messianic" ethics, open to the future, not eliminate the need for such political management? I will return to these knotty questions in my analysis in Chap. 7 of the relation between politics and technology.

4.8 Summary

Derrida's understanding of time proceeds, on the one hand, from Heidegger's thinking about the event and the expression "*es gibt,*" and, on the other, from the deconstruction of Husserl's transcendental thinking on the basis of the temporalization of consciousness. Derrida replaces Husserl's transcendental logic with another mode of thinking, grounded in *différance* and *avenir*. On this view, time is manifested in the form of *inscription* in the space that is opened up as *writing* and in the form of an originary temporal *synthesis*, whose movement is maintained by the inscription of writing itself and which forms the condition of possibility for all meaningful experience of time. Hence this arche-writing—or *différance*—replaces Husserl's transcendental logic as a framework for producing meaning.

At the same time, Derrida connects *différance* as a framework for producing meaning with the concept of *avenir*, which is to say the other, the completely other, that suddenly—*as if incidentally*—emerges from an uncertain and unforeseeable future. However, *l'avenir* should be regarded not as something that will occur at some future moment in time but rather as something that intervenes with messianic force to interrupt the homogeneous flow of time here and now. This messianic "something" that comes from without is what Derrida calls *the gift*. As he sees it, the pure gift, if such a thing exists, cannot be repaid. It cannot, without losing its status as a gift, be a stage in an economic circuit; it must be absolutely incompatible with the return gift, with economic exchange, with the economy as a whole, which is always represented by a *circle*. For this reason it is also absolutely incompatible with the economy of time because time in the traditional sense is likewise represented by a circle. When such a gift, which can also be unwelcome and destructive, disrupts the homogeneous circle of time, the result is a complex and qualitatively more heterogeneous time, a time that cannot be understood as linear but that instead appears to be bent, a time that cannot be understood in terms of "before" and "after." *The time is out of joint*.

Dislocated time becomes dislocated precisely because time is produced artificially. It is an *artifact* produced by a teletechnological apparatus that necessarily

[63] Stiegler, "Återerinring och minnesstöd. Platon som den förste att tänka proletariatet [Remembrance and memory support. Platon as the first to think the proletariat]", pp. 234 and 236 (my trans.).

creates pseudo-events through the editing, distorting, and reformatting of "information" that is frequently incomplete or detached from its context. According to Derrida, a pseudo-event produced in this way is inscribed within the very centre of the event structure even as it is necessarily differs from the originary event itself. It is the task of deconstructive thinking not to conflate appearance with reality but instead to analyse the pseudo-event by also being sensitive to the originary event.

Yet Derrida's understanding of time is more radical than this. For him, not only these pseudo-events but *all* events are technologically mediated. If all events must be inscribed as traces *in space*, then these traces, these representations of technology, must constitute time by means of the movement of *différance*. Derrida's time is hardly compatible with the view that time is constituted on the basis of an ultimate foundation, whether that foundation is comprised by a transcendental subjectivity, an existential temporality, or an empirical technicity. On the contrary, Derrida's "messianic" time is open to the possibility of the impossible. Ultimately, this is deconstruction's quest and its passion: to remain open to an uncertain and unforeseeable future—to *l'avenir*.

References

Derrida, J. (1995a). *Mal d'archive: Une impression freudienne*. Galilée.

Derrida, J. (1995b). *Archive Fever: a Freudian impression*. (trans. E. Prenowitz). The University of Chicago Press.

Derrida, J. (1994). *Specters of Marx: The state of the debt, the work of mourning, and the new international*. (trans. P. Kamuf). Routledge.

Derrida, J. (1993a). *Given time 1: Counterfeit money*. (trans. P. Kamuf). University of Chicago Press.

Derrida, J. (1993b). *Aporias*. (trans. T. Dutoit). Stanford University Press.

Derrida, J. (1992). Force of law: The "Mystical Foundation of Authority". In D. Cornell, M. Rosenfeld, & D. G. Carlson (Eds.), *Deconstruction and the possibility of justice*. (trans. M Quaintance). Routledge.

Derrida, J. (1991). *Donner le temps 1: La fausse monnaie*. Galilée.

Derrida, J. (1987). *Psyché, invention de l'autre*. Galilée.

Derrida, J. (1982a). Différance. In *Margins of philosophy*. (trans. A. Bass) (pp. 1–27). The University of Chicago Press.

Derrida, J. (1982b). Ousia and Grammè. In *Margins of philosophy*. (trans. A. Bass) (pp. 29–67). The University of Chicago Press.

Derrida, J. (1982c). *Margins of philosophy*. (trans. A. Bass). The University of Chicago Press.

Derrida, J. (1978a). *Writing and difference*. (trans. A. Bass), The University of Chicago Press.

Derrida, J. (1978b). *Edmund Husserl's origin of geometry: An introduction*. (trans. J.P. Leavey, Jr). Harvester Press.

Derrida, J. (1976). *Of grammatology*. (trans. G. C. Spivak). The John Hopkins University Press.

Derrida, J. (1972a). *Marges – de la philosophie*. Minuit.

Derrida, J. (1972b). *Position*. Minuit.

Derrida, J. (1972c). Ousia et Grammē. In *Marge – de la philosophie* (pp. 31–78). Minuit.

Derrida, J. (1967a). *De la grammatologie*. Minuit.

Derrida, J. (1967b). *L'écriture et la difference*. Seuil.

Derrida, J., & Stiegler, B. (1996). *Échographies de la télévision*. Galilée.

Derrida, J., & Stiegler, B. (2002). *Echographies of television: Filmed interviews*. (trans. J. Bajorek). Polity Press.

Hägglund, M. (2008). *Radical atheism: Derrida and the time of life*. Stanford University Press.

Rosa, H. (2013). *Social acceleration: A new theory of modernity*. (trans. J. Trejo-Mathys). Columbia University Press.

Stiegler, B. (1998). *Technics and time, 1: The fault of epimetheus*. (trans. R. Beardsworth & G. Collins). Stanford University Press.

Stiegler, B. (2011). Återerinring och minnesstöd. Platon som den förste att tänka proletariatet [Anamnesis and Hypomnesis. Plato as the first thinker of the proletarianisation] (trans. L. Dahlberg). In L. Dahlberg & H. Ruin (Eds.), *Fenomenologi, teknik and medialitet*. Södertörn University.

Wood, D. (1989). *The deconstruction of time*. Humanities Press.

Chapter 5
Humanity's Technicity

5.1 Introduction

In philosophy of technology circles, there is currently an intense debate about the possibility and desirability of "enhancing" human beings by means of modern technology. The accelerating convergence of information technology, biotechnology, nanotechnology, and cognitive science—the so-called NBIC technologies—may lead to radically new applications, such as brain-computer interfaces, recreated human organs, and perhaps even beings that are part-human, part-technology.

The issue of *human enhancement*, that is to say, the improvement of human beings in various ways by NBIC and other technologies, is strongly associated with the philosophical movement known as posthumanism. Posthumanists such as N. Katherine Hayles, Donna Haraway, and Peter Sloterdijk often foreground the advantages of using modern technology to enhance human beings. In recent years, however, the so-called transhumanist movement has begun to gain ground. This utopian movement, one of whose most prominent exponents is Oxford professor Nick Bostrom, wants to go a step further: not content with enhancing human beings, it instead wants to create an entirely new, transhuman form of life. Its goal is to fundamentally change human nature by using modern technology, with a view to eliminating ageing and greatly increasing human beings' intellectual, physiological, and psychological capacities. Quite simply, it wants to create a successor to human beings. Opponents of this transhumanist credo argue in defence of human dignity; they oppose all attempts to manipulate the human genome and claim that technologies that do not respect the limits of (the transcendental) subject and that undermine the foundations of its autonomy represent a potential threat to its freedom, independence, and individuality. As Andrew Feenberg has recently observed, human beings are infinitely complex beings whom we have barely even begun to understand. In actual fact, we know extremely little about what a human being really is.

B. Sjöstrand, *Derrida and Technology: Life, Politics, and Religion*,
https://doi.org/10.1007/978-3-030-83407-4_5

In a lecture titled "The Aforementioned So-Called Human Genome" that Derrida gave at an interdisciplinary conference in 1992, he takes up the ancient question about humanity, about the essence of a human being: *what is a human being?*[1] There are, he argues, really two ways of formulating the question: are human beings essentially beings that have knowledge of themselves? With a capacity to improve themselves? Or is a human being ultimately an "I" that asks questions about *ethics*, *freedom*, and *responsibility*? What we are actually dealing with here is two competing definitions of the same being. On the one hand, humans can be defined as beings who have knowledge of themselves—scientific and technological—and who explore the consequences of that knowledge. On the other hand, there are beings who at least ask questions about ethics, freedom, and responsibility, things about which they not only lack but *must* lack that kind of knowledge. Not because such knowledge does not exist but because not having it is a requirement for being able to make responsible decisions. If I know what it is that I am supposed to do, that is, if I possess the requisite knowledge of what I should do before I make a decision, I am not, strictly speaking, making a decision. Rather, I am simply applying a programme or rule. Derrida also argues that technological-scientific knowledge itself presupposes a shortage or lack of knowledge. In other words, this lack of knowledge seems to be a precondition not only for behaving in an ethically responsible way but also for technological-scientific development as such.

Derrida is deeply ambivalent about the accelerating pace of technological and scientific development and feels torn between two "contradictory feelings." The first is a sense of tragedy that is unique in human history: an unprecedented advance in knowledge is now associated with immense power, not only technological and scientific power but also technological-economic and technological-political power. Based on knowledge, this power is essentially concentrated to a few superpowers in the rich world. This hegemony of near-total knowledge is a source of worry for Derrida, an intensely apocalyptic fear that he evokes dramatically in the following passage:

> One has the impression (and I am describing my first impression, that of a *tragic, apocalyptic pathos*) … that the risk that is run at this unique moment in the history of humanity is the risk of new crimes being committed against humanity and not only, if I can say this, against millions of real human beings as was the case, but a crime such that a sorcerer's apprentice who was very cunning, the author of potential genetic manipulation, might in the future commit or supply the means for committing—against man, against the very humanity of man, no longer against millions of representatives of real humanity but against the essence itself of humanity, against an idea, an essence, a figure of the human race, represented this time by a countless number of beings and generations to come.[2]

Derrida evidently imagines that such an apocalypse, or at least the possibility of one, would take different forms at what he calls "this unheard-of moment." "Are we not," he asks himself, "at this unheard-of moment … in a situation where a terrible suspicion sets the stage for a dramatic judgment whose apocalyptic proportions

[1] Derrida's lecture is collected in *Negotiations*, pp. 199–214. The conference was organized by Association Descartes and took place in December 1992.

[2] Derrida, "The Aforementioned So-Called Human Genome", pp. 207–8, emphasis added.

begin to resemble a last judgment?"[3] This judgment, whose principal charge would be *crimes against humanity*, would be directed not only against technological-scientific research and all its consumers but also against all the politicians, citizens, philosophers, and others who challenge or even abolish the classic concepts of *consciousness*, *freedom*, and *responsibility*. For this reason, Derrida argues, it is essential that leading philosophers ask more questions about humanity's essence, about the concept of the human and how it should be understood: "The question, *What is man*? could no longer wait as it seems to have done formerly."[4]

Alongside this sense of "fatal precipitation," Derrida experiences a diametrically opposite feeling: a sense of calm. He adduces no fewer than seven reasons for this relative calm:

1. at the end of the day, these questions are not especially new and have been debated throughout all of history;
2. "the *human genome* is not man," i.e. being *able* to map the human genome is not the same as being able to manipulate it;
3. biologists, geneticists, decision-makers, and financers are on the whole "pretty vigilant" and have enough historical memory to put up effective barriers against the temptation to apply eugenicist principles and eliminate "inferior" human beings or create super-humans[5];
4. there is now a strong legal and political awareness of these problems that is also being refined, and binding laws and conventions are being regularly adopted with a view to regulating these enormous advances in knowledge;
5. we can and must resist all forms of ignorance and disinformation;
6. between mystical obscurantism and positivistic scientism, there lies another way: "even if we do not really know what the human is, and even if it is not the object of knowledge … even if we do not know who we are … the human is no more in danger today … than it was yesterday or the day before"; and

[3] Ibid., pp. 208–9.

[4] Ibid., p. 209.

[5] Derrida's relative trust in biologists and geneticists, decision-makers and financers, warrants closer scrutiny. After the German program of racial hygiene, which ended in the extermination of millions of people whom the Nazis considered "inferior", all programmes of racial and hereditary hygiene naturally received official condemnation. Post-humanist and transhumanist optimists instead put their hopes in the spectacular advances made in science and technology, such as genetic engineering, nanotechnology, medical technology, computer technology, and human-computer symbiosis. One area of technology that has once again become highly topical is artificial intelligence (AI). The potential risks of AI have been debated among futurists for decades but only in recent years has the tone become more urgent. After a period of temporary ups and downs, AI research has gained real momentum in the past five years, thanks in large part to improved methods and resources for managing massive amounts of data. Eventually, AI will perhaps be able to learn on its own, and one day its intelligence will perhaps surpass that of humans, leading to an "intelligence explosion." Transhumanists such as Hans Moravec and Ray Kurzweil predict that this will happen in a few decades. At the same time, respected scientists and innovators such as Shane Legg and Elon Musk warn that AI represents a major threat, perhaps even the greatest threat, to humanity in this century.

7. finally, let us not forget that these rapid developments of knowledge are making possible spectacular advances in medicine.[6]

What is a human? The question is important not only because it is urgent but also because it is intimately bound up with the question concerning technology, Heidegger's *Die Frage nach der Technik*. For example, if we extend our definition to include some kind of teleological anthropology, according to which humans are made in God's image, there are consequences for how we view the human-technology relation. Not that teleological anthropology in principle has to be technology's opposite term, but rather that humans as the image of God are, intrinsically, radically other to technology. For this reason, teleological anthropology is opposed to transhumanism. They may have the same objective, namely a new kind of human being, but their means are quite different. Transhumanism rests on technology, teleological anthropology on spirituality.

If, on the other hand, we extend our definition to include a particular kind of philosophical anthropology, the result is a different human-technology relation. For instance, the German biologist and philosopher Helmuth Plessner regards humans as by their nature *artificial*. They are and have always been cyborgs, that is to say, human-machine hybrids. From this view of human beings as "artificial," it does not seem unreasonable in principle to argue for the desirability of changing humans by means of various artificial interventions.

In the next section, I will show that Derrida follows another path, moving beyond these variants of teleological and philosophical anthropology. I begin by summarizing his view that technology is always already inscribed into human life as well as into life in general, after which I discuss the way in which he develops the classical phenomenological concept of *subjectivity*.

5.2 Technology's Inscription in Life

One common way of understanding what technology is involves counterposing it to nature. For Derrida, however, the concepts of nature and technology, *physis* and *techne*, are not polar opposites; rather, nature is always already inseparably bound up with technology. More specifically, this intertwining governs the conventional phrases "humans and technology" and "life and technology." For Derrida, life—human life as much as life in general—is always already intertwined with technology. It is thus not only that technology is not the opposite term of life, it is from the earliest possible moment *inscribed* in it.

In his introduction to *De la grammatologie* (1967), Derrida lays the foundations for a new science of writing by introducing the general concept of *gramme*,

[6] Ibid., p. 209–10.

meaning "written mark."[7] He connects this concept with the concept of *programme*, noting that biologists in his lifetime began to refer to writing and *programme* in connection with "[t]he most elementary processes of information within the living cell."[8] He also claims that the governing *programme* in the emergent field of cybernetics is writing. Not only the most elementary processes of genetic inscription but even the programmes of cybernetics are thus examples of writing, of *gramme*. And the concept of *gramme* cannot be dispensed with: "If the theory of cybernetics is by itself to oust all metaphysical concepts—including the concepts of soul, of life, of value, of choice, of memory—which until recently served to separate the machine from man, it must conserve the notion of writing, trace, *grammè*."[9] According to Derrida, these three closely-related concepts are so-called quasi-transcendentals, which, like transcendentals, constitute the conditions of possibility for knowledge and meaning albeit with this difference: that the experiences they make possible exceed their own conditions of possibility. They are not metaphysical concepts, argues Derrida, and resist definition within metaphysics because they themselves are the origin and conditions of possibility for all metaphysical concepts. Nevertheless, he chooses to provisionally describe this trace as the movement in *différance* that "opens appearance [*l'apparaître*] and signification [by] articulating the living upon the nonliving in general"[10] and *gramme* as that trace's "element."[11]

It is no coincidence that Derrida connects *gramme* with the programmes of genetics and cybernetics in this way, since it relates to the adoption of a new "language" by molecular biologists at around that time.[12] At the same time as Derrida was writing down his ideas about *gramme* in 1965, the older technical vocabulary of molecular biology was in the process of being replaced by a more modern terminology of "coding," "decoding," "message," "transmission," and so forth. The young Derrida followed this development closely and took theoretical inspiration from the new sciences of information technology and cybernetics.

The 1960s was an exciting transition period, not least because biology and genetics had hitherto been treated as inferior sciences that lacked the theoretical weight of mathematics and physics. It is easy to imagine the excitement of biologists and geneticists at that time when digital DNA technology and the mathematical correlation of nucleic and amino acids transformed molecular biology, more or less overnight, into a respected discipline to be counted among the hard sciences.

In 1962, Francis Crick, James Watson, and Maurice Wilkins shared the Nobel Prize for the discovery of DNA's double-helix structure, which in turn paved the way for the deciphering of genetic code. French biologists and geneticists also took

[7] Derrida uses "*grammè*"—from the Greek word *grammē* (line, trace) and reminiscent of *gramma* (letter)—as a neologism related to *différance*. As with *différance*, I have chosen not to translate it.

[8] Derrida, *De la grammatologie*, p. 19.

[9] Ibid.

[10] Ibid, p. 95.

[11] Ibid, pp. 19–20.

[12] Liu, *The Freudian Robot*, pp. 61–2.

part in this race to solve "the riddle of life" by building advanced cybernetic models that they described using the technology-inspired language of information theory. What had previously been termed "structural specificity" was now called genetic "information." The various functions of the DNA molecule were likewise now described in terms of "copying," "transcription," and "transmission," and even the mechanisms regulating the transfer of genetic information were redesignated *écriture de l'hérédité*—hereditary writing.[13] In recognition of their work, French molecular biologists François Jacob, André Lwoff, and Jacques Monod received the Nobel Prize in Medicine in 1965.

Derrida's ideas about *gramme* in *De la grammatologie* are more easily understood when viewed in relation to a larger debate taking place in France at that time about the analogy between writing and living cells. Philosophers such as Roman Jakobson and Claude Lévi-Strauss participated intensively in the debate, as did Foucault, who wrote *inter alia* a long and positive review in *Le Monde* of François Jacob's book *La logique du vivant* (1976) (translated as *The Logic of Life*). In it, Jacob coined the concept of "programme"—as a deliberate analogy to the programmes by which electronic computers operate—to describe the structure of the molecule regulating living organisms.[14] The living cell was increasingly beginning to resemble a machine, and the machine a living cell.

Joining this debate, Derrida wrote the first chapter of *De la grammatologie* as a review of three books, including a newly published two-volume work by the French paleo-anthropologist André Leroi-Gourhan titled *Le geste et la parole* (1965) (translated into English as *Gesture and Speech*). This extraordinarily ambitious study, which came to exert a profound influence on Derrida, spans an enormous period of time, from the beginnings of life to twentieth-century information society.[15] According to Leroi-Gourhan, who was deeply interested in technology and philosophy, the discovery of the early hominid *Zinjanthropus boisei* in 1959 created an entirely new picture of how human beings originate, a picture that emphasized the fundamental importance of technology and writing for the development of human intelligence. *Zinjanthropus*—which lived in Central Africa about 1.8 million years ago—walked upright on two legs and used advanced tools, despite having an extremely small brain (the size of a chimpanzee's). By walking upright, *Zinjanthropus* freed their hands for the use of tools; conversely, the use of tools freed the lower jaw for the development of speech and language, which in turn freed the brain for symbolic thinking and intelligence.

Leroi-Gourhan's critique of the prevailing view of humanity's origins takes the form of a new, literally inverted anthropology. Traditionally, humans have been thought of as having first developed a larger brain (three or four times that of a chimpanzee) before using it to create human culture. Leroi-Gourhan upends this

[13] Ibid., p. 62.

[14] Ibid., p. 64.

[15] Some of the data presented in *Le geste et la parole* is out of date in consequence of subsequent archaeological research. However, the new findings do not invalidate Derrida's philosophical arguments and conclusions—quite the reverse.

logic: we acquired greater brain capacity by moving away from genetically determined behaviour patterns. What Zinjanthropus teaches us, Leroi-Gourhan argues, is that our big brains and their capacity for consciousness, intelligence, and symbolic thought were not the precondition for tools; on the contrary, tools were what made big brains possible. Moreover, for Leroi-Gourhan, the entire history of humanity, from the development of the human skeleton all the way to today's media and information society, is a history of the gradual unleashing or externalization of human potential by means of technology. Last but not least, Leroi-Gourhan is extremely interested in the "programmes" mentioned earlier, particularly the operating programmes of electronic brains, and he argues that these programmes are developing so quickly that they are blurring the boundary between human and machine.

In a curious passage in Part One of *De la grammatologie*, Derrida takes up and radicalizes Leroi-Gourhan's ideas. He here defines *différance* as life. According to Derrida, *différance* must be understood as a *programme* for life, encompassing the whole chain of the genetic inscription of life, via the emergence of human beings, up to and beyond alphabetic writing. To make his line of argument comprehensible, I will quote the passage at length:

> Leroi-Gourhan no longer describes the unity of man and the human adventure thus by the simple possibility of the graphie in general; rather as a stage or an articulation in the history of life-of what I have called *différance*-as the history of the *gramme*. Instead of having recourse to the concepts that habitually serve to distinguish man from other living beings (instinct and intelligence, absence or presence of speech, of society, of economy, etc.), the notion of program is invoked. It must of course be understood in the cybernetic sense, but cybernetics is itself intelligible only in terms of a history of the possibilities of the trace as the unity of a double movement of protention and retention. This movement goes far beyond the possibilities of the "intentional consciousness." It is an emergence that makes the gramme appear *as such* (that is to say according to a new structure of nonpresence) and undoubtedly makes possible the emergence of the systems of writing in the narrow sense. Since "genetic inscription" and the "short programmatic chains" regulating the behavior of the amoeba or the annelid up to the passage beyond alphabetic writing to the orders of the logos and of a certain *homo sapiens*, the possibility of the gramme structures the movement of its history according to rigorously original levels, types, and rhythms. But one cannot think them without the most general concept of the *gramme*. That is irreducible and impregnable. If the expression ventured by Leroi-Gourhan is accepted, one could speak of a "liberation of memory," of an exteriorization always already begun but always larger than the trace which, beginning from the elementary programs of so-called "instinctive" behavior up to the constitution of electronic card-indexes and reading machines, enlarges *différance* and the possibility of putting in reserve: it at once and in the same movement constitutes and effaces so-called conscious subjectivity, its logos, and its theological attributes.[16]

Let us consider what Derrida is really trying to say in this unnecessarily obscure and inaccessible passage. He begins by questioning traditional philosophy's sharp distinction between human beings and other living creatures. His reading of Leroi-Gourhan is based on the view that human beings are merely one "stage" in the larger history of life, that is to say, the history of *gramme*. Thereafter he connects *gramme* with the concept of programme, which can be found on both sides of the

[16] Derrida, *De la grammatologie*, p. 125–6.

demarcation line. In other words, he is emphasizing the *continuity* of the concept of programme, from genetic inscription, via alphabetic writing, all the way to "electronic card registers" and "machine readers," i.e. systems of technological inscription. This continuity thus has its basis in the possibility of *gramme*, which should be understood as an extremely general programme that precedes all individual types of programme, whether genetic or non-genetic. And, Derrida argues, even if the "specific levels, types, and rhythms" of the programmes must be taken into consideration: "we cannot imagine them without using an extremely general conception of *gramme*. This is irreducible and incontrovertible."

Let us pause here for a moment and reflect on these striking claims. Is it really an "irreducible and incontrovertible" fact that human beings are merely one stage in the larger history of life, or that there is a continuity in the "concept of programme" from genetic inscription to alphabetic writing and beyond? Is it really self-evident that *gramme* as programme precedes any individual kind of programme, and that we cannot imagine these without "an extremely general conception of *gramme*"?

Bernard Stiegler approaches the question of humanity's origins very differently from Derrida. In Part One, Chap. 3 of *La technique et le temps* (1994), he discusses what he calls the "invention of the human." For Stiegler, humans are an "invention" and even an "embryonic fabrication or conception."[17] And the key to the discovery of the human is precisely the role that Leroi-Gourhan ascribes to *technology* in human evolution. For Leroi-Gourhan, human evolution is not only the evolution of a biological being but also the evolution of technological aids—such as writing and tools—without which humanity would never have arisen. For Stiegler, human history is "a history of technics" in which "it is *technē* that invents the human, not the human who invents the technical."[18] In his view, humanity's discovery of technology, the "exteriorization" of humanity in technology, is a unique event in the history of life. Something *happens*, something *is begun* at the very moment when externalization first takes place. *That* something radically new begins at that moment, that a passage *occurs*, a "passage of the cortex into flint" is, for Stiegler, irreducible and incontrovertible. This is the point at which Stiegler differs from Derrida.[19]

As we have seen, Derrida's reading of Leroi-Gourhan presupposes that the externalization of the human in technology is a *stage* in the larger history of life. Stiegler regards this step, this passage "from the genetic to the nongenetic," as "the appearance of a new type of *gramme* and/or program." At the same time, however, Derrida insists upon an extremely general concept of *gramme* and upon the *continuity* in the concept of programme, such that this concept can be found on both sides of the demarcation line. For this reason, Stiegler regards Derrida's position as inconsistent: *gramme* constitutes both the history of life in general and a specific stage in the history of life, which Stiegler associates with the discovery of the human.

---

[17] Stiegler, *Technics and Time, 1*, p. 135.

[18] Ibid., p. 141.

[19] Ibid., p. 141.

Stiegler seems to regard nature and technology as so strictly distinct from each other that it is only after this passage that a specific living creature—human beings—are connected by technology. And yet, as Ben Roberts and others have pointed out, there is no reason to assume that Derrida endorses this view.[20] Quite the opposite. When Derrida refers to *gramme* as being an extremely general concept that encompasses not only the "non-genetic" but also "genetic inscription," he does so on the basis of his conviction that nature and technology are not each other's polar opposites but in fact always already inseparably interwoven.

Derrida's conviction that this extremely general conception of *gramme* constitutes the conditions of possibility for both genetic and non-genetic inscription seems reasonable when we consider that genetic inscription "*does indeed* articulate the living upon the non-living in general": the DNA molecule in a living organism binds it with its non-living ancestors to precisely the same degree as those ancestors' written and technological legacy; genetic codes preserve in living organisms the legacy of the non-living in a way that is analogous to (though of course not identical with) writing and technology. Moreover, genetic inscription is not a one-time occurrence but a repetitive, mechanical process: from its earliest beginnings and in every instant, life is "a process of self-replacement, the handling-down of life is a *mechanike*, a form of technics."[21] This does not mean that Derrida considers life *as such* to be a form of technology, but from a certain perspective it must be described in technological terms. As he writes, immediately following the passage cited above, in relation to Leroi-Gourhan's summary in *Writing and Speech* of the relation between hand and face:

> Leroi-Gourhan describes the slow transformation of manual motricity which frees the audio-phonic system for speech, and the glance and the hand for writing. In all these descriptions, it is difficult to avoid the mechanist, technicist, and teleological language at the very moment when it is precisely a question of retrieving the origin and the possibility of movement, of the machine, of the *techne* ... In fact, it is not difficult, it is essentially impossible.[22]

Derrida is thus arguing that human beings, from a certain perspective, must be described in "technological" terms. At the same time, he claims that it is impossible

[20] See, for example, Ben Roberts, "Stiegler Reading Derrida", p. 5.

[21] Derrida, "Nietzsche and the Machine", p. 244.

[22] Derrida, *De la grammatologie*, p. 126. Derrida returns to the problematic of the hand in his late text *Le Toucher, Jean-Luc Nancy*. In an exemplary deconstruction of the philosophy of touch, he highlights precisely the technicity of the hand. Derrida begins by emphasizing how significant the hand is for the phenomenological tradition, not least for Husserl's thinking as it is exemplified in his *Ideas II*, which is entitled *Phenomenological Investigations into the Constitution of Reality in its Totality*. For Husserl, the hand is the very symbol of our existence as free, volitional, spontaneous, and knowing beings. At the same time, he criticizes Husserl's understanding of the hand, which he sees as characterized by pure metaphysics of presence. Instead, inspired by the anthropology of Leroi-Gourhan, he describes the human hand as an *originary supplement*—as the first tool or the first prosthesis. As he notes, it is impossible to dissociate the history of the hand from the history of technology. For a more detailed discussion of the problematic of the hand in *Le Toucher*, see Arthur Bradley, *Originary Technicity*, pp. 113–122.

to disregard the origin and possibility of this "technology": the extremely general conception of *gramme*. But is it really true that this *gramme* precedes technology in the way that Derrida claims? Wasn't it the other way round, i.e. that the metaphorical use of the concept of programme by biologists and cyberneticists inspired his own concept of *gramme*? The subtle wordplay between *gramme* and *programme* in French represents a slight modification of Leroi-Gourhan's discussion of electronic brains' "*programmes,*" which Derrida rewrites in the singular: *the programme*. Derrida's idea is that *gramme* as *programme* constitutes the genetic and non-genetic programmes, but the immediate context in which *De la grammatologie* was written actually indicates the opposite.

Whatever the case, it is clear that Derrida has tried to establish a fundamental relation between the various programmes of technology and genetics on the one hand, and, on the other, *gramme* and the trace as movement in *différance*. Does this mean that intentional consciousness is constituted technologically, as Stiegler claims? Not at all. The trace "at once and in the same movement constitutes and effaces so-called conscious subjectivity, its logos, and its theological attributes." Even if humans must in some regards be described in terms of technology, they cannot be defined as technological-biological beings, nor even as a purely transcendental consciousness. Instead, it is the quasi-transcendental character of the trace that constitutes (and puts under erasure) human beings' consciousness of themselves, society, and the world.

If humans cannot be reduced to pure biology, and if they have also been stripped of their transcendental qualities, what are they, then? In Part Two of *De la grammatologie*, titled "Nature, Culture, Writing," Derrida develops his new logic—*the logic of the supplement*—in relation to a reading of Rousseau's texts, according to which culture, language, writing, and technology are aporetically necessary *supplements* to nature. In other words, humans cannot have existed *prior* to their supplements but, rather, must have appeared *retroactively* as a result of this originary supplementarity. As Derrida, in polemic with Rousseau, writes: "Man allows himself to be announced to himself after the fact of supplementarity, which is thus not an attribute—accidental or essential—of man." In other words, the supplement does not add itself in any simple fashion following the emergence of humans, as a kind of foreign body; rather, this foreign—and dangerous—supplement is from the beginning active and located in the seemingly ideal interior of body and soul. But what *is* supplementarity, then? Nothing, claims Derrida: "Supplementarity *is nothing*, neither a presence nor an absence, is neither a substance nor an essence of man. It is precisely the play of presence and absence, the opening of this play that no metaphysical or ontological concept can comprehend."[23] If human identity is therefore always already intimately bound up with its supplements in a way that cannot be comprehended, this insight presumably means that Derrida is questioning the project of creating a rigorous science of the human, a *human* science. Is this perhaps why he writes, just before the long passage quoted above, that grammatology "ought

[23] Derrida, *De la grammatologie*, p. 347.

not to be one of *the sciences of man*, because it asks first, as its characteristic question, the question of the *name of man*"?[24]

To summarize: *De la grammatologie* radically challenges the prevailing view that life and technology are each other's polar opposites. The history of life, from the amoeba to *Homo sapiens* is indissolubly intertwined with what Derrida calls the history of *gramme*. The continuity between genetic inscription, alphabetic writing, and information storage in electronic archives that *gramme* makes possible appear to be fundamental for him. However, this does not mean that he is an adherent of a biological reductionism that views humans as purely technological-biological beings. In fact, he is every bit as opposed to biological reductionism as he is to metaphysical subjectivism.

At the same time, he argues that we cannot shake off (*se démarquer de*) biologism in its genetic form without embracing some form of metaphysical subjectivism. A strategy of presenting biologism *in opposition to* metaphysical subjectivism inevitably leads to our being forced to choose that which is "the least grave."[25] The alternative strategy chosen by Derrida involves developing the classical phenomenological concept of subjectivity into what I will be calling "scriptual subjectivity."

5.3 Scriptual Subjectivity

In several of his earlier works, particularly *Speech and Phenomena*, Derrida discusses the classical concepts of subject and subjectivity. The starting point for this discussion is his view that the Western philosophical tradition has always been characterized by a fundamental prejudice—namely the privileging of the concept of "presence"—a prejudice that he calls "the metaphysics of presence" and that constitutes the fundamental basis for classical understandings of subjectivity, from Descartes, via Kant, to Husserl. As Derrida sees it, this tradition of a metaphysics of presence reaches its absolute zenith in Husserl's phenomenology. Husserl and the tradition have always interpreted the meaning of being as presence, an interpretation that takes two forms: (1) something *is* insofar as it appears to a subject as a present object in sensory observation or as pure objectivity; (2) a subject *is* only insofar as it is self-present, i.e. present to itself in the immediacy of the psychical act. This very immediacy, this idea of the subject's self-presence, is what Derrida calls into question.

Derrida treats the issue of subjectivity as self-presence in greatest depth in the fifth chapter of *Speech and Phenomena*, in which he carries out a deconstructive analysis of Husserl's understanding of temporal consciousness as it appears in his so-called Lectures on Time, *Vorlesungen zur Phänomenologie des inneren*

[24] Ibid., p. 124.

[25] Derrida, *De l'esprit*, pp. 65–66. Derrida is here referring to Nazism's two opposing ideological roots: biologism and glorification of the (German) spirit.

Zeitbewußtseins (1905). Derrida begins by emphasizing the seriousness of what is at stake, namely that the *simultaneous* self-presence and immediacy of the present moment are of fundamental significance for the entire phenomenological project. We might even go so far as to say that "the general possibility of a primordial perception or intuition" stands or falls on "such a perception or intuition of self by self in presence."[26] As Derrida continues:

> if a certain concept of the "now," of the present as punctuality of the instant ..., is a myth, a spatial or mechanical metaphor, an inherited metaphysical concept, or all that at once, and if the present of self-presence is not simple, if it is constituted in a primordial and irreducible synthesis, then the whole of Husserl's argumentation is threatened in its very principle.[27]

It is not difficult to agree with Derrida that this concept of punctuality—the present as a point on an imagined time-line—which goes back to Aristotle's *Physics*, is problematic, to put it mildly. The fact is that Husserl not only admits this, he also adapts his entire project to the idea that "the living present" must have a certain extension in time. If this were not the case, if consciousness were only conscious of that which is given in a punctual now, if a stream of consciousness thus consisted of a series of isolated now-points strung out as if on a pearl necklace, we would not be able to experience temporal objects, let alone their movements. But at the same time as Husserl insists upon the duration of the living present, he clings to the punctual now's self-identity as the "principle of principles"—as the "source point" for certain knowledge.[28] It is precisely here, at the very heart of Husserl's phenomenological discourse, that Derrida discovers an inherent contradiction. Even though Husserl insists continually on the punctual now as the "primal form" (*Urform*) of consciousness, Derrida points out that Husserl in his Lectures on Time "prevents us from speaking about a simple self-identity, *im selben Augenblick.*"[29]

In an attempt to get around this contradiction, Husserl describes an object in terms of three inseparable psychical acts: (1) the perception of the object's now-phase, (2) the primary memory of the object's preceding phrase ("retention"); and (3) the primary expectation of what will take place ("protention"). When Derrida analyses Husserl's discussion of perception, retention, and protention in greater depth, he reaches a surprising conclusion:

> One then sees quickly that the presence of the perceived present can appear as such only inasmuch as it is continuously compounded with a nonpresence and nonperception, with primary memory and expectation (retention and protention). These nonperceptions are neither added to, nor do they occasionally accompany, the actually perceived now; they are essentially and indispensably involved in its possibility.[30]

In other words: if presence is to have any *meaning*, it must inevitably and essentially be followed by non-presence. This presence and non-presence presuppose

[26] Derrida, *Speech and Phenomena*, p. 59.

[27] Ibid., p. 60.

[28] Ibid., p. 61.

[29] Ibid., p. 68.

[30] Ibid., p. 64.

each other, they are always already *interwoven*, they constitute a continuity. If we acknowledge this continuity, Derrida says, we create a space for *the other* in self-identity, for "the nonpresence and nonevidence …in the *blink of the instant*. … This alterity is in fact the condition for presence."[31] In so doing, he has *implicitly* introduced the very condition of possibility for the self into Husserl's discourse as the movement in *différance*.

If movement in *différance* involves a continuity between presence and non-presence, between perception and retention, it may be asked whether there is a *radical* difference between retention and *re-presentation*, between primary and secondary memory (recollection, imagination, etc.). Derrida agrees with Husserl that there is a difference but that it is not radical. Rather, there are different modifications of non-perception. As Derrida explains:

> Without reducing the abyss which may indeed separate retention from re-presentation, without hiding the fact that the problem of their relationship is none other than that of the history of "life" and of life's becoming conscious, we should be able to say a priori that their common root—the possibility of re-petition in its most general form, that is, the constitution of a trace in the most universal sense—is a possibility which not only must inhabit the pure actuality of the now but must constitute it through the very movement of difference it introduces. Such a trace is—if we can employ this language without immediately contradicting it or crossing it out as we proceed— more "primordial" than what is phenomenologically primordial.[32]

This key passage requires careful reading. In the first place, we can note that "the possibility of re-petition"—non-presence's return in presence—is what Derrida calls a "*trace*." This trace does not merely "inhabit" the now, it "constitutes" it *a priori*. Constitution thus occurs *a priori* when the trace introduces *movement in différance* into "the pure actuality of the now." As a result, the trace is more "'primordial'" than all the modalities of phenomenological presence. And Derrida describes the system of traces as the "arche-writing" that precedes and determines metaphysical presence.

When Derrida describes the system of traces as "arche-writing," he is using an image that emphasizes the double nature of writing, partly as writing in the traditional sense (for example, the writing on this page) and partly as a subjectless condition of possibility for knowledge and meaning. It is important to emphasize that in the latter case the concept of writing is not limited to traditional writing. The attributes found in traditional writing, according to Derrida can be generalized and hold true "not only for all orders of 'signs' and for all languages in general but moreover, beyond semio-linguistic communication, for the entire field of what philosophy would call experience."[33] All human experience is thus made possible by this originary writing, this "arche-writing." The human subject as consciousness can therefore no longer be described as the centre of being; rather, it must be understood as an effect within a system that is no longer a system of presence but a system of traces that is open to changes in the world around it. *Différance* expresses this

[31] Ibid., p. 65.

[32] Ibid., p. 65.

[33] Derrida, "Signature Event Context", p. 9.

openness because the production of meaning is only possible if presence is related to something other than itself in such a way that it can include the trace of the past and always already allow itself to be contaminated by the future.

Yet Derrida's use of writing in this figurative sense in the place of the classical subject does not eliminate the need for a psyche. According to Derrida, this psyche, what Freud calls "the psychical apparatus," should be thought of as a "writing machine" (*machine d'écriture*) that is continually inscribing and erasing traces of the world around it. Collectively, these traces constitute a system of relations in a way that cannot be reduced to the classical subject's privileging of presence:

> The "subject" of writing does not exist if we mean by that some sovereign solitude of the author. The subject of writing is a system of relations between strata: the Mystic Pad, the psyche, society, the world. Within that scene, on that stage, the punctual simplicity of the classical subject is not to be found.[34]

Accordingly, we can no longer regard the subject as the expression of a sovereign inner life whose essence is unaffected by the external context. This external context is always already related to the psychical through the trace, through what Derrida calls "the possibility of re-petition." He places particular emphasis on the importance of *repetition* for subjective experience, even "insisting" that the possibility of the sign functioning in a normal context is conditional upon its capacity to break with this context and create new contexts and thence new meanings.[35] In this way, the sign can only be repeated or, as Derrida often puts it, "be iterated" infinitely, resulting in the generation of endlessly new meanings. Iteration does not mean repetition of a given meaning, therefore, because there is nothing that decides that a sign should have a particular given meaning. In other words, it is not possible for there to be a pure form of meaning that corresponds to the sign's "essence"—a claim that he extends to all human experience. All experience must be interpreted.

Derrida's iterability principle[36] may seem to undermine the very possibility of subjectivity as it is understood in classical epistemology and ethical theory. But this is not his aim. Instead, he wants to *redefine* the subject by describing it as a collection of general texts that in principle has no boundary. An important consequence of this redefinition is that subjective experience is always intertwined with various technologies for storing and disseminating text. Text-based technologies such as the Internet determine the character and content of our experience to a very high degree, because the way in which they archive texts is (historically) new: "what is no longer archived in the same way is no longer lived in the same way."[37]

In other words, Derrida sees experience as technologically modifiable. However, this does not mean that he views the subject through some kind of techno-scientific lens. The fact that he spreads out and externalizes the subject's functions does not

---

[34] Derrida, *L'écriture et la difference*, p. 335.

[35] Derrida, "Signature Event Context", p. 12.

[36] Derrida points out in "Signature Event Context" p. 7 that *iter* comes from *itara*, which means *other* in Sanskrit. He thereby connects iterability with alterity.

[37] Derrida, *Mal d'archive*, p. 37.

mean that he wants to describe the possibility of experience in techno-scientific terms. He continually points out that the possibility of experience is an "effect" of the general text's quasi-transcendental character. It is thus not technology but *différance* that creates phenomenological experience. And, as we recall, *différance* and the other quasi-transcendentals are of course "more 'primordial' than what is phenomenologically primordial."

5.4 Summary

Derrida's ambivalence about the accelerating developments in technoscience, his simultaneous apocalyptic pathos and relative calm, seems legitimate. It *warns* us, it supplies us with contradictory signals that we can neither ignore or close our eyes to. A conceivable development in the spirit of posthumanism or transhumanism would therefore require *vigilance* but also *openness* to an unknown future that can never be predicted.

In turn, Derrida argues, this kind of openness requires that the question *What is a human?* be addressed with renewed energy. For him, this question is strongly associated with the question concerning technology. Whether we choose to understand humans as beings who have technological-scientific knowledge of themselves and who accept the consequences of this knowledge, or whether we regard the essence of the human as an *ego* that asks fundamental questions about *ethics*, *freedom*, and *responsibility*, it follows that human beings *a priori* require *supplements* in the form of culture, language, writing, and technology.

Derrida's position in this matter differs from Plessner's "artificial" anthropology or Stiegler's technologically determined view of human beings in that Derrida understands human life not as technology but as always already *contaminated* with technology. Life is not technology but nor is it the opposite of technology; rather, technology is inscribed in life from the very start. The appearance of humanity is thus "merely" one stage in the history of life in general. There is, Derrida argues, a certain continuity that ranges from genetic inscription, through alphabetic writing, to the storage of information in electronic archives in the present day.

Surprisingly, Derrida's philosophy of technology finds expression in his redefinition of the classical subject. Indeed, his development of Husserl's phenomenological discourse means that the subject can no longer be understood as the centre of being but must instead be described as an effect within a system of writing that is open to the technological world around it. For him, the psyche or the psychical apparatus seem to be a "writing machine" that continually inscribes and erases traces of that surrounding technological world. This means that we can no longer represent the psychical apparatus as being *in opposition to* a non-psychical apparatus. In fact, these two apparatuses are becoming increasingly alike, and there is a clear analogy between our own internal memory and today's complex systems of inscribing and storing almost unlimited amounts of information in external memory. It is for this reason that Derrida in *Freud et la scène de l'écriture* (translated as

Freud and the Scene of Writing) can define the concept of technology, not as an apparatus, but as the relation between two apparatuses: the psychical and the non-psychical apparatuses—between life and death.

References

Bradley, A. (2011). *Original technicity: The theory of technology from Marx to Derrida*. Palgrave Macmillan.

Derrida, J. (2002). Nietzsche and the machine., The aforementioned so-called human genome. In E. Rottenberg (Ed.), *Negotiations: Inventions and interviews, 1971-2001*. Stanford University Press.

Derrida, J. (2000). *Le toucher, Jean-Luc Nancy*. Galilée.

Derrida, J. (1995a). *Archive Fever: a Freudian impression*. (trans. E. Prenowitz). The University of Chicago Press.

Derrida, J. (1995b). *Mal d'archive: Une impression freudienne*. Paris.

Derrida, J. (1989). *Of spirit: Heidegger and the question*. (trans. G. Bennington, R. Bowlby). The University of Chicago Press.

Derrida, J. (1988). Signature event context. In G. Graff (Ed.), *Limited Inc.*. (trans. S. Weber, J. Mehlman). Northwestern University Press.

Derrida, J. (1987). *De l'esprit: Heidegger et la question*. Galilée.

Derrida, J. (1978). *Writing and difference*. (trans. A. Bass). The University of Chicago Press.

Derrida, J. (1976). *Of grammatology*. (trans. G. C. Spivak). The John Hopkins University Press.

Derrida, J. (1973). *Speech and phenomena, and other essays on Husserl's theory of signs*. (trans. D. B. Allison). Northwestern University Press.

Derrida, J. (1967a). *De la grammatologie*. Minuit.

Derrida, J. (1967b). *L'écriture et la difference*. Seuil.

Leroi-Gourhan, A. (1993). *Gestures and speech*. (trans. Anna Bostock Berger). MIT Press.

Leroi-Gourhan, A. (1964). *Le geste et la parole*. Albin Michel.

Liu, L. (2010). *The Freudian robot: Digital media and the future of the Unconscious*. The University of Chicago Press.

Roberts, B. (2007). Stiegler reading Derrida: The prosthesis of deconstruction in technics. In *Postmodern Culture*, Vol.16, No7.

Stiegler, B. (1998). *Technics and time, 1: The fault of Epimetheus*. (trans. R. Beardsworth & G. Collins). Stanford University Press.

Chapter 6
Ethics and Technology

Deconstruction is—if it is—an ethics ... If you call this an ethics of affirmation, it implies that you are attentive to otherness, to the alterity of the other, to something new and other—
Jacques Derrida, 1992

6.1 Introduction

Beginning in the early 1990s, Derrida often made it clear that the ethical is never absent in deconstruction, indeed, that deconstruction *is* an ethics. The deconstructive thinking that in his previous works may have seemed like theoretical analyses of language and text should actually be understood as ethical thinking. Although a number of commentators have come to refer to an "ethical turn" in Derrida's later writing, he has himself denied that this is the case.[1] Rather, it is a question of the ethical implications of moral, social, and political questions becoming increasingly explicit since the late 1980s. As I will try to show, this ethical dimension in his thinking is closely associated with *technology*. This chapter will examine the nature of this connection and how we ought to understand it.

There is probably no other philosopher in the phenomenological tradition who has examined the technological dimension of ethics as thoroughly as Derrida. Neither Husserl, nor Heidegger, nor Levinas discuss technology in ethical terms. Levinas hardly mentions technology at all apart from a two-page essay in *Difficile liberté* titled "Heidegger, Gagarine et Nous" ("Heidegger, Gagarin, and Us"), in which he criticizes Heidegger's philosophy of technology. In my view, it is important to highlight Derrida's unique line of argument about technology's inscription in ethics, not least because it has an important contribution to make to the discussion of technology's role in the global democratization process. Let us begin by briefly considering Derrida's relation to ethics.

Initially, Derrida was considered as anything but a philosopher who wrote about ethics. In his first writings from early 1960s, ethical questions are treated in an

[1] Derrida, *Voyous*, p. 64.

B. Sjöstrand, *Derrida and Technology: Life, Politics, and Religion*,
https://doi.org/10.1007/978-3-030-83407-4_6

"oblique," unthematic fashion, both suggestively and evasively. Although it sometimes seems as though he is not interested in ethics, he in fact engages with ethical problematics in a highly committed and fundamental way.

Derrida has been criticized by numerous commentators on the grounds that deconstruction lacks a stable foundation for ethical or moral position-taking, or, perhaps even worse, that it seeks to put an end to ethical thinking in general by replacing it with a kind of ethical relativism or moral nihilism that is without principles and that is unable to serve as the basis for political action. Other critics, such as Michel Rosenfeld, have argued that the strong emphasis on *différance* at the price of identity and rationality leads to indecisiveness, an inability to choose sides, and, with it, political relativism.[2] Like Derrida himself, I argue that this is a misunderstanding. Deconstruction does not advocate a particular politics; it advocates what it does, namely, questioning the current basis for political thinking and political action. These foundations are also limits that must be exceeded if anything new is to emerge. As we will see, Derrida's writings have engaged with ethical questions in this spirit from the very start. Deconstruction is anything but indecisive; on the contrary, it claims to be able to change the world through responsible decision-taking.

As far back as one of his first texts, *De la grammatologie* (1967), Derrida summarizes the ways in which ethics—the ethical—is closely intertwined with language and writing. What interests him, however, is not some undifferentiated understanding of ethics but an understanding that is closely connected with the rest of his thinking. He is interested in the *relation to the other*.[3] What exactly does this mean? It means that the self is nothing without another person or thing with which it has a fundamental, constitutive relation. This is the central ethical idea, which has deep roots in Emmanuel Levinas's ethical theories, that Derrida develops and varies in many different ways throughout his entire oeuvre.

Published in 1964, *Violence et métaphysique* is perhaps the work in which Derrida's ethical thinking appears most clearly. But his particular view of ethics is also evident in a series of other texts, not only those that deal with obviously ethical themes such as friendship (*Politiques de l'amitié*), hospitality (*De l'hospitalité*), illegal immigration (*Negotiations*), or ethnocentrism (*De la grammatologie*), but also those that address how philosophy has consistently ignored particular definite problems such as marginality (*Marges—de la philosophie*), supplement (*De la grammatologie*), and many other problems that philosophy throughout its history has neglected in its quest for truth, identity, and meaning. All in all, I would argue that Derrida's philosophical orientation *as a whole* is ethical in a particular sense of the word that I want to try to narrow down more precisely.

Derrida's particular view of ethics can be largely traced back to Emmanuel Levinas. Derrida was the first to really confront Levinas's thinking when he used his 1964 book *Violence et métaphysique* to demonstrate a number of contradictions in Levinas's magnus opus *Totalité et infini*. Levinas's ethical thinking thereafter came

[2] Rosenfeld, "Derrida's Ethical Turn and America", p. 845.

[3] Derrida, *De la grammatologie,* p. 69.

to exert a strong and uninterrupted influence on the development of his own thought, particularly from the 1980s and on. Derrida's thinking and Levinas's run parallel to each other in many regards, but there are also dissonances and the dialogue between them is both complicated and ambiguous. Without going into this dialogue further, I will nonetheless argue, as have many of Derrida's commentators, notably Roberto Bernasconi and Simon Critchley, that Derrida's deconstructive thinking is ethical in the sense that Levinas understands the term.

In *The Ethics of Deconstruction* (1992), Simon Critchley argues that "Derridian deconstruction can, and indeed should, be understood as an ethical demand, provided that ethics is understood in the particular sense given to it in the work of Emmanuel Levinas."[4] For Levinas, "ethics" means an unqualified responsibility for another human, for the Other. Levinas's key idea is that this ethics is *the first philosophy*, that ethics precedes ontology. The ethical relation to the Other thus cannot be thematized or conceptualized; it cannot be reduced to pure understanding. There is something in the Other—a dimension of inaccessibility, of exteriority, of what Levinas calls *alterity*—that I can never understand.

Derrida himself characterizes Levinas's ethics, not as a *theory* of ethics, but as an ethics of ethics, the condition of possibility for all ethics. This means that it cannot give rise to a particular form of ethics in the traditional sense without being disloyal to itself.[5] For Levinas, ethics is thus an ethics that precedes moral questions and, to an even greater degree, moral laws; instead, it is an *originary* ethical experience from which moral questions can be derived.[6] This experience is an experience of the *alterity* of the Other, an encounter with the face of the Other as it is revealed to me; it is an experience that calls into question my own self as a subject possessing absolute power. By questioning my dominance as subject, the face of the Other liberates me from identity with myself, thereby laying a foundation for the ethical.

Although Derrida firmly rejects the notion that ethics was the first philosophy, he takes up and modifies Levinas's discourse on the relation to the Other.[7] However, he views the relation as involving not merely another human being but also *the other*: other textual interpretations, other ways of looking at things, other cultures, other genders, other times (particularly the future), other languages, other traditions—and other technologies. When I therefore claim that Derrida's texts have a deep undercurrent of ethics, it is because of their uninterrupted engagement with the other as person, people, and thing, etc. This engagement involves both a duty and a responsibility to alterity.[8] Indeed, it could be claimed that virtually all of Derrida's thinking

[4] Critchley, *The Ethics of Deconstruction*, p. xiii.

[5] Derrida, *Violence et métaphysique*, p. 164.

[6] Critchley, *The Ethics of Deconstruction*, p. 3.

[7] Derrida's choice of a lower-case "a" in expression such as "the other" is highly deliberate and significant. Unlike Levinas, he almost always avoids using capitalized letters, in order to underscore the non-religious aspect of the expression.

[8] The concept of "responsibility" in Derrida is inspired not only by Levinas's ethics but also by the Christian tradition. In his reading of Jan Patočka's *Heretical Essays in the Philosophy of History* in *Donner la mort* (1992), he tries to clarify the difference between how Platonic philosophy and

can be summarized under the rubric of *alterity*. Considered in this way, his writing, from the very beginning of his oeuvre, can be seen as having forged a path of its own, driven by an ethical imperative of opening up to the world, to the other, the entirely other.

For Derrida, ethics is openness to the entirely other. This openness involves an *experience* of alterity. Such experience is always aporetical, since it presupposes a thinking that does not reduce the alterity of the other to pure identity. This aporetical experience is what Derrida explicitly defines as *deconstruction*.[9] The fact that deconstruction is an experience of alterity means that it is driven by an ethical imperative and a responsibility for the entirely other, the unconditional, what he also calls "the non-negotiable." The work of deconstruction does not consist of negotiating about the "negotiable." Rather, it is "prescribed by ethical duty itself" to "negotiate the non-negotiable," to negotiate about every seemingly unsolvable problem that the world is wrestling with (such as terrorism).[10]

The fact that deconstruction is an aporetical experience means that the aporia is neither a pure antinomy or a dialectical opposition in the Hegelian or Marxist sense or even a "transcendental illusion" of the Kantian variety, but rather an *interminable* experience. According to Derrida, an aporetical experience must necessarily have endurance if we are to be able to make ethically responsible decisions.[11] This understanding of aporia reveals a kind of *temporal* structure in deconstruction; it is fundamentally an aporetical way of thinking *in time*. Yet it is primarily engaged with the future, not the present.

What does Derrida mean by future? In the first place, he distinguishes carefully between the two French terms *avenir* and *futur*. The word *futur* refers to that which is expected, predictable, or known in advance—known with the same certainty that usually characterizes our knowledge of the past. *Le futur* refers to the way in which we normally talk about the future, i.e. as we perceive it when it has already occurred. Much of our knowledge of the future can be predicted more or less exactly, such as the fact that sun will rise tomorrow. According to Derrida, however, French differentiates between this predictable future, *futur*, and *avenir*, that which is to come—*à venir*—that which comes to us from without, from a place that we can neither know about nor foresee. To use another of his metaphors, *futur* is what we can see on the

---

Christianity view responsibility, by investigating the ways in which they each understand death. Yet Derrida's interest in the responsibility structure of Platonic philosophy is mostly instrumental in relation to his interest in Christianity's unique structure of responsibility. For Platonic philosophy, death is an integral component in the philosophical care for one's self, which is in turn intimately associated with the care of the city state or *polis*. For Christians, the gift of death is always already immanent in the gift of life, in the new, authentic life that triumphs over death and that lasts forever. The transition from the Platonic structure of responsibility to the Christian is visible, above all, in the importance ascribed to *the heart*. God's gaze sees into the human heart, judging and rewarding it from within. The divine gaze that lights up the human heart is an event that opens up the infinite love that sacrifices itself and shows charity to the other (pp. 40, 93).

[9] Derrida, *Psyché: Inventions de l'autre*, p. 27.

[10] Derrida, "Ethics and Politics", p. 304.

[11] Derrida, *Apories*, p. 35.

horizon—horizontally, while *avenir* strikes our head vertically like a bolt from the blue.

For Derrida, what comes to us suddenly like a bolt from the blue is an *event*. And for an event to be an event, it must be *im-possible*. If it were possible, it could be traced back to the present or the past. For Derrida, however, the event is something that comes from without, from some other place—from the future that is to come—*à venir*—and not from that which has already happened. A true event is therefore "im-possible."

This somewhat paradoxical formulation demands an explanation. On the face of it, Derrida's statement seems counterintuitive, particularly since it seems to run contrary to the *possibility* of an originary ethics. But when he refers to the event as "im-possible," he does not mean that it cannot happen; rather, he means that it happens beyond the *possibility* of the subject foreseeing it, beyond its horizon of expectations, beyond its capacity to anticipate it. We can often anticipate what is going to happen, he says, but "the incalculable happens [*arrive*]."[12] The im-possible is not nothing; it is *openness* to the event. Derrida counterposes the subject's capacity to figure out what is going to happen—its "I know"—to "the impossible, of what must remain (in a non-negative fashion) foreign to the order of my possibilities."[13] The im-possible exists, he claims, as "the possibility of this impossible."[14]

The event is *the other, the entirely other* that we can never foresee. If it could be foreseen, it would not be an event. Whether it rains tonight is not an event in Derrida's sense of the word because we know what rain is. A true event is completely unforeseeable; it comes suddenly and entirely unexpectedly, it does not wait, it does not give us time to consider what is happening. It intervenes and imposes itself. It takes me by surprise, it causes a shock, a wound, or even a trauma. Because it takes me by surprise, it is impossible for me to understand or take in what is actually happening. Such an event is *absolutely unique*; it is impossible to speak of it in general terms without losing its uniqueness.

Although such absolutely unique events *can* be experienced live and in real time, it is increasingly experienced by means of *technological mediation*.[15] "Today, more than ever before," says Derrida, the event is produced and presented by a teletechnological "media apparatus,"[16] that includes not only the media in the form of press, radio, and television but mobile systems, digital archiving systems, and a global infrastructure in the form of industries, experts, agreements, and modes of working. This teletechnological media apparatus creates the event, produces it, interprets it, and perhaps even reinterprets it, in accordance with its own interests: the event becomes an *artifact*.

---

[12] Derrida, *A Taste for the Secret*, p. 61.

[13] Derrida, *Philosophy in a Time of Terror*, p. 134.

[14] Ibid., p. 115.

[15] Strictly speaking, the event can never be experienced as pure spectatorship: ("'live' and 'real time' are never pure: they do not give us intuition or transparency, a perception stripped of interpretation or technical intervention". (*Echographies de la télévision*, p. 13).

[16] Derrida & Stiegler, *Échographies de la télévision*, p. 11.

It is "necessary," argues Derrida, to deconstruct this *artifactuality*, this technological and artificial production. It is necessary because deconstruction requires taking responsibility for the underlying event, the "real" event such as it has actually occurred.[17] Taking responsibility here means ensuring that the real event is not erased from memory, that the dead—all of those individual deaths—are not forgotten when the spotlight is switched off and the images have gone from the TV screen.

It is also necessary—indeed, even a duty—to be open to "the newest technical events, toward the most surprising sophistications of technics, such as they are happening or will happen to us, coming upon us from the future [*l'à-venir*]." Why is it a duty to be open to future events? Because, Derrida explains, these events dramatically challenge the concept of *democracy*. This concept, which we have inherited from antiquity, is determined and delimited in the modern era by the boundaries of the nation-state, by a particular territory, by everything that we think we understand by the word "citizenship," based on blood ties and territory. Derrida is here arguing that these technological events will entail a "de-territorialization" of the political. We may even need to imagine democracy beyond the "limit" of the political. Being open to these technological events is necessary because "we are given this imperative—to think the political beyond the political, as it were, or the democratic beyond democracy—by technics, concretely, every day—both as a threat and as a chance."[18] If deconstruction is an ethics, that ethics is necessarily bound up with technology.

But ethics is not loosely associated with technology, it is "in an essential way and, in any case, to a large extent" (*de façon essentielle, et en tous cas pour une large part*) bound up with the transformative power of media and teletechnology.[19] Although certain events in recent years, such as the fall of the Berlin Wall and the end of apartheid in South Africa, were foreseeable, prior to their occurrence not even the most well-informed experts could have predicted when they would take place. The moment for the event was determined in large part by the transformative power of technology, particularly television. As Derrida sees it, the increasingly rapid transformation of technology itself is an important driving force on the road towards a world characterized by cross-border democracy and greater justice: "I believe that this technical transformation—of the telephone, of the fax machine, of television, e-mail and the Internet—will have done more for what is called "democratization," even in countries in the East, than all the discourses on behalf of human rights, more than all the presentations of models in whose name this democratization was able to get started."[20]

---

[17] Ibid., p.13.

[18] Ibid., p. 76. The idea that ethics is closely associated with technology is not new. It can be found in the work of philosophers such as Jacques Ellul and Hans Jonas, even if their views on ethics differed partly from Derrida's. Hans Jonas also spoke about an ethical imperative—the imperative of responsibility—according to which new technology "demands a corresponding ethics of foresight and responsibility". But where Jonas regarded this technology primarily as a threat and a risk, Derrida saw it as not only an unparalleled threat but also an opportunity for politics and democracy.

[19] Ibid., p. 82.

[20] Ibid.

Derrida is thus arguing that technology *in practice* can bring about greater democracy and justice. Yet this does not mean that technology always promotes these values; it can just as easily work against them. It can be associated with new, destructive ways of using existing technology. Disturbing examples include the use of commercial passenger jets as deadly weapons, the potential use of chemical, biological, and nuclear weapons of mass destruction, and the threat of cyberwarfare intended to quickly knock out global communication systems and archives.

To illustrate the connection between ethics and technology, I devote the remainder of this chapter to a discussion of three events of world-historical importance: the electronic revolution; total nuclear war (as a hypothesis), and global terrorism. Derrida himself has not examined these issues in systematic fashion but instead considered them thematically in different contexts that I will be grouping together under three headings.

6.2 The Electronic Revolution

Derrida has written extensively about the electronic revolution, a revolution that has changed human life. It is hard not to view it as one of the most revolutionary global changes that the world has undergone since the industrial revolution. It is undeniably a world-historical event, which has essentially taken place during a relatively short period of time, namely the last few decades. During this short space of time, a growing stream of new technological inventions has seen the light of day. For example, the first digital mobile phones appeared during the 1970s, the first personal computers a decade later, services like email and the World Wide Web in the 1990s, and e-books, e-readers, and the whole range of social media just a few years ago. Derrida describes this cultural revolution, still ongoing and involving a transition from a paper culture to an electronic culture, as a "geo-techno-logical shock" and an "earthquake" followed by successive "aftershocks."[21]

Derrida took an interest in this "geo-techno-logical shock" throughout his writing career. In *La dissémination* (1972), *Marges—de la philosophie* (1972), *Glas* (1974), *La carte postal* (1980), *La monolingualisme de l'autre* (1996), and other books, he discusses the concept of *paper* and what happens when a world of paper is replaced by apparatuses or machines for electronic inscription. From the very start, in fact, he has been deeply engaged with the *subject* of paper:

> I have the impression … that I have never had any other *subject*: basically, paper, paper, paper. … I have always written, and even spoken, *on* paper: on the subject of paper, an actual paper, and with paper in mind.[22]

Two works in particular are central to the issue of paper and its relation to the electronic revolution: *Mal d'archive* (1995) and *Papier Machine* (2001). In the

[21] Derrida, *Mal d'archive*, p. 33–4.

[22] Derrida, *Papier Machine*, p. 239.

introduction to *Papier Machine*, Derrida wonders what exactly is happening. What *happens* between paper and machine? What new experience of *the event*? What happens to the event? What happens to its archive when *the world of paper*, which has existed for hundreds of years, is subordinated to all these new machines for virtualization? Can a virtual event really exist? A virtual archive?[23] Derrida searches anxiously for a direction for his thinking at a moment when pronouncements about the end of a historical period defined by paper are being bandied about. To begin with, what does the word *paper* mean? And what should we understand as a *machine*? Are we talking about two nouns with the same status (paper *and* machine, paper *or* machine)?

Derrida reminds us that the French word *papier-machine* literally refers to the *material*, the *sheet* that serves as the *basis* for a typewriter's writing and for the printing, reproduction, and archiving of what is produced by computers and similar apparatuses (copier paper). But paper is naturally not only writing or copier paper but also other kinds of paper: wrapping paper, wallpaper, cigarette paper, toilet paper, and so on. Paper can thus be used for anything. It is therefore necessary, Derrida argues, to differentiate between different kinds of paper. First, there is paper that represents "the condition of a priceless archive, the body of an irreplaceable copy, a letter or painting, an *absolutely unique event*" (emphasis added). Second, there is the kind of paper that represents the basis for commercial products; and, third, there is simple waste paper.[24]

What is happening right now with the two first kinds of paper, namely that we are experiencing a rapid and simultaneously unforeseeable reduction—a withdrawal, a "de-paperization," even, if we can speak of it thus—reminds us that "paper has a history that is brief but complex, a technological or material history."[25] Paper is a *technology* that has dominated for a particular period in human history. When we write by hand, therefore, with pen and paper, we are not living in a pre-technological time; there are and have always been apparatuses or machines, comprehensive reproduction and mechanical iterability. It is therefore no legitimate to counterpose manuscript to mechanical writing, a pre-technological craft as the opposite of technology.[26]

Paper is not the only technology, however. Before the invention of paper, people used parchment made from animal hide. Yet parchment is not paper: it cannot be used as a surface for printing on. These two materials therefore belong to two separate technological periods and systems of inscription. Derrida now asks himself whether there is a general structure concealed *behind* both these technologies, a *quasi-transcendental* paper:

> *Behind* these specific determinations (the support made of skin, or the paper, and other ones too), beyond or before them, might there be a sort of general, even quasi-transcendental

[23] Ibid., p. 10.

[24] Ibid., p. 242.

[25] Ibid., p. 241.

[26] Ibid., p. 152.

> structure? A structure both superficial, actually that of a surface, and yet also profound enough and sensitive enough to receive or retain impressions? When we say "paper," for example, are we naming the empirical body that bears this conventional name? … Or are we by the same token designating this "quasi-transcendental paper," whose function could be guaranteed by any other "body" or "surface," provided that it shared some characteristics with "paper" in the strict sense of the word (corporality, extension in space, the capacity to receive impressions, and so on)?[27]

Derrida is here arguing that these different meanings of the word *paper* are always superimposed; they are, so to speak, piled on top of each other in so complicated a fashion that the question "What is paper?" does not have a straightforward answer.[28] However, the fact that Derrida neither can nor wants to define what paper *is* does not mean that he has succumbed to the temptation of reducing "the technological event," the invention of all new electronic and digital apparatuses for multimedia, to "being merely a *development* of paper."[29] It would be a serious mistake, he argues, not to understand that the new—virtualization, the enormous change in tempo, the immediate and transparent globalization—is something else entirely. It is *something radically other*: it represents nothing less than a *technological step* in human externalization.[30]

This technological step in human history gives us much food for thought as regards our relationship to new technology. One problem that Derrida highlights is our dependence upon a host of multimedia apparatuses. Even if people know roughly how they are supposed to use a particular apparatus, such as a computer, they do not have the faintest idea about its internal workings or the protocols governing its operation. This secrecy without mysticism elicits a feeling that we are dealing with the soul of *someone Other*. At stake here is a kind of face-to-face relation to "a benign or evil genius," an invisible and omnipotent observer who instantly understands our instructions and responds to them *without delay*. At the same time, however, it is as if the Other, the "Unconscious-Other," is speaking to us in a *mechanical* way, without sympathy. A power cut or a careless movement at the computer can delete several hours' work in the blink of an eye. Advantages such as greater speed, greater freedom, and more powerful tools are quickly outweighed by a feeling of subservience, insecurity, and alienation.[31]

Derrida also identifies other problems with paperless technology: an exponential increase and deregulation of trade in copyrights and patents; the spread of invisible power structures (particularly the dominance of the English language and Anglo-American culture); the potential loss of the centuries-old experience of being able to identify oneself by means of various kinds of paper, something that can create a feeling of being without an identity:

---

27 Ibid., p. 253–4.

28 Ibid., p. 254.

29 Ibid., p.247.

30 Ibid., p.259. For a longer discussion of technology and human externalization, see Chap. 5.

31 Ibid., p.156.

> "The paper is mine"; "Paper is a self or ego"; "Paper is me." Paper often became the place of the self's appropriation of itself, then of becoming a subject in law. As a result, in losing this tangible body of paper, we have the feeling that we are losing that which protected that subjectivity itself.[32]

We need only think of our ever-nagging anxiety about losing the identity papers that we have learned to rely upon. But this anxiety is nothing compared to the catastrophe entailed for one's self-image by the loss of one's identity papers—by becoming *sans papiers*, paperless. Identity papers of various kinds will therefore always exist, Derrida predicts. They are a kind of *spectre of paper*; they have a sacred force, that of the law.[33]

At the same time as he points out all the problems associated with paperless culture, Derrida places equal emphasis on the potential benefits of that culture. Personally, he confesses to feeling as though he is drowning under *too much paper*, something that makes him irritable. He wants to save the world *from* paper![34] And he dreams of living a paperless life, which sounds to him like the definition of "real life, of the living part of life."[35] He also highlights the ecological benefits associated with a reduction in the use of paper (for instance, that fewer trees are sacrificed to paper mills).[36] Although he thinks it risky to draw up a balance sheet of threats and opportunities, he argues that the potential "benefit" of the new technology is undeniable: it facilitates and speeds up "storage, archiving, communication, and debates beyond social and national frontiers; a hyperactive circulation of ideas, images, and voices; democratization, homogenization, and universalization; immediate or transparent 'globalization.'"[37]

Derrida's involvement with the subject of *paper* and with the issue of what happens when a world of paper is replaced by machines for electronic inscription is an example of ethical thought, if by ethics we mean openness to the entirely other. Electronic and digital technology precisely represents the entirely other, the radically new in human development: virtualization, the enormous change in tempo, an immediate and transparent globalization, and deterritorialization. For Derrida,

---

[32] Ibid., p.259.

[33] Ibid.

[34] Derrida loved computers and TV almost as much as books: "I love every form of the book and I see no reason to give this love up. But I also love—this is the fate of my generation, of just this one generation— the computer and the TV. And I like writing with a pen just as much, sometimes just as little, as writing with a typewriter—mechanical or electric—or on a computer." (*Papier Machine*, p.29). And again: "I began by writing with a pen … I wrote [then] more and more "straight onto" the machine: first the mechanical typewriter; then the electric typewriter, in 1979; then finally the computer, around 1986 or 1987. I can't do without it any more now, this little Mac, especially when I'm working at home" (*Papier Machine*, p. 153).

[35] Ibid., p. 272.

[36] Ibid., p. 269.

[37] Ibid., p. 258. These threats and opportunities seem to be as topical today as they were in 2001 when Derrida wrote his book: globalization, a millennial buzzword, is still defined by rapid global streams of information, ideas, culture, capital, people, goods, and services.

ethical engagement, which has its basis in this new technology, demands a new way of thinking about politics—and democracy—beyond territorial boundaries.

As an example, let us consider the *book* of the future. What will happen to the book? Will it disappear in the form that we know it? Will we even be able to use our libraries if they no longer contain books? Will they be turned into work stations for the reading and writing of electronic texts? And will literature and philosophy—as we know them—be able to survive without paper books?

Derrida discusses these and similar questions in a chapter titled "Le livre à venir" ("The Book to Come") in *Papier Machine*. As we have already seen, the French phrase *à venir* refers to that which comes to us from without, from a place that is unforeseeable, from some place *other* than what we expect. Derrida is invoking this sense when he begins by proposing that the expression "Le livre à venir" can mean at least three things:

1. That the book as such has—or does not have—a future;
2. That if it has a future, the book to come will no longer be what it was; and
3. That we are awaiting or hoping for an *other* book, a book to come that will transfigure or even rescue the book from the shipwreck that is happening at present.[38]

Derrida alludes to "shipwreck" because, quite apart from the possible death of the book, he finds it painful to watch the paper-based book losing ground to other media and ways of reading and writing. For earlier generations, and to some extent his own, the book represented something spiritual, indeed, almost sacred. The book as a Work (with a capital "W") by a consecrated author has traditionally enjoyed a status that is almost comparable to that of the codex in a previous age, with its manuscripts and autograph manuscripts. We must, Derrida argues, reflect on this "quasi sacrality" that has shaped the entire technological history of inscription and archiving. It seems as though every stage in a technological revolution, by means of a long history of enlightenment and reason, desacralizes, democratizes, and secularizes the book; by the same measure, each of these stages seems also to be followed by a sacred or religious "reinvestment."[39] This process will always necessarily give rise to an opposing force that seeks to realize what Derrida calls a *resacralization*, a reinvestment in every aspect of the book (above all, when prices rise as a result of reduced availability). There will always be enthusiasts who remain passionate about the paper book and gather around it like a kind of secular priesthood: producers, interpreters, decision-makers, and others. There will therefore always be a tension between *dispersion* and *reinvestment*. This tension will define the very form of the book to come. What is happening today is.

> *on the one hand* … the irreversible dispersion of this total codex (not its disappearance but its marginalization or secondarization … *on the other hand*, a constant reinvestment in the book project, in the book of the world or the world book, in the absolute book … the new space of writing and reading in electronic writing … It re-creates the temptation that is figured by the World Wide Web as the ubiquitous Book finally reconstituted, the book of

[38] Derrida, *Papier Machine*, pp. 20–21.

[39] Ibid., p. 24.

> God, the great book of Nature, or the World Book finally achieved in its onto-theological dream.[40]

With this almost Hegelian formulation, Derrida conjures up two "fantasies" about the fate of the book, two extreme and eschatological images of the end of the book: its death or its *telos*—its fulfillment (*accomplissement*). His point is that we cannot *foresee* what will happen and that we must therefore keep an open mind about what might happen. Admittedly, he is convinced that the paper book will not disappear, that it will survive in some form and co-exist with the new technologies. But it is crucial that we do not commit ourselves in advance to any particular view, whether pessimistic or optimistic, reactionary or progressive. We must, argues Derrida, resist catastrophic pessimism, a pessimism that predicts the book's death. This is entirely separate from the fact that such pessimism is the result of being drawn into futile resistance to an inevitable technological development whose benefits are, moreover, "so obvious, not just in terms of efficiency and economy but also ethically and politically."[41] But we must also watch out for a progressive—and sometimes "romantic"—optimism that is primed to connect the new technologies to the myth of the absolute book that has no material basis, to the myth of universal transparency, instant communication, and a complete absence of limits or any form of surveillance or control, and to the myth of the global democratic village. This is the optimism of a new *Aufklärung* that is ready to sacrifice, indeed, even to burn, all of the old books and libraries in the name of Enlightenment and of Reason.

As regards the book's *future*, Derrida warns us against succumbing to the temptation of adopting *in advance* a pessimistic or optimistic stance. We must embrace what comes *no matter what it may be*. "[T]he affirmation of the future *to come* is not a positive thesis," he writes: "It is nothing other than the affirmation itself, the "yes," insofar as it is the condition of all promises or of all hope, of all awaiting, of all performativity, of all opening toward the future."[42] At the present moment in time, to say "yes" to the future that is to come—to the entirely other—and that represents an opportunity as well as a threat, is to act ethically.

This ethical responsibility also recurs in Derrida's treatment of *the archive*. The question of the archive, which we briefly introduced at the end of Chap. 3, recurs continually in Derrida's writing, such that we might tentatively claim that, for him, the archive represents the centre around which both paper culture and electronic culture are entirely organized.[43] In *Mal d'archive*, which is subtitled *Une impression freudienne* and which was originally given as a lecture to an international conference on psychoanalysis in London in 1994, Derrida deconstructs the concepts of

[40] Ibid., p. 27.

[41] Ibid., p. 29–30.

[42] Derrida, *Mal d'archive*, p. 109.

[43] There are numerous passages in Derrida's writing to support this statement, e.g.: "As techno-science, science, in its very movement, can only consist in a transformation of the techniques of archivization, of printing, of inscription, of reproduction, of formalization, of ciphering, and of translating marks" (*Mal d'archive*, p. 31).

archive and archiving. On the basis of an etymological analysis of the Greek word *arkhe*, he finds that the archive is the physical, historical, or ontological place where a document or law is preserved and thus the place from which political power is wielded. From its very beginnings, the archive was closely associated with the political power and violence that both inscribe and conserve the law. This *archival violence* inscribes the law at the same time as it compels people to respect it. The archive has the "force of law."[44]

Yet the question of the archive is not limited to the political sphere but is also highly relevant within the sphere of the private. One thinker who took a strong interest in the question of the archive was Sigmund Freud. In the first chapter of *Mal d'archive*, Derrida refers to *Civilization and Its Discontents* (1929–30), a book in which Freud underscores the technological archive's importance for psychoanalysis. For example, medical records were first archived using paper and ink, then printing presses and printer's ink. As the volume of archived material grew prodigiously, a mountain of paper and ink piled up, seemingly to no purpose. Freud asks what the point is of all this archiving of narratives that basically everyone knows about: why should we invest in so much paper, ink, letters, and work? He argued that these enormous archives are in fact worthless and should be destroyed. Partly as a consequence of this, he felt compelled to introduce a new thesis into his theory of psychoanalysis: we must always remain attentive to "the possibility of a radical perversion, indeed, a diabolical death drive, an aggression or a destruction drive."[45] This drive, to which Freud here gives three names, strives *to destroy the archive*; it is what Derrida calls *anarchic*, it exercises *archival violence*.[46] It seeks to eliminate not only the archive as such but also to remove any trace of its own presence. Archival violence completely eliminates the entire technological archive and its accompanying apparatus of documentation; it is simply the locus of the total obliteration of its own memory. It is this death drive—in glaring contrast to Freud's famous pleasure principle—that Derrida terms "archive fever":

> Another economy [than the pleasure principle] is thus at work, the transaction between this death drive and the pleasure principle, between Thanatos and Eros ... The death drive is not a principle. It even threatens every principality ... every archival desire. It is what we will call, later on, *le mal d'archive*, archive fever."[47]

This *archive fever*, this drive whose purpose is to destroy the archive, stands in opposition to the desire to preserve, what Derrida calls the *archive drive*. We all have a powerful need to preserve and search in archives, and some people even have an almost passionate relationship to the archive, an irresistible desire to search back in time for the origin; they nurture a feeling of homesickness, a nostalgic longing for the place where everything once began. While this archive drive naturally has its spatial and temporal limitations, the threat represented by the destructive force of

---

[44] Ibid., p. 20.

[45] Ibid., p. 23.

[46] Ibid., p. 24–5.

[47] Ibid., p. 27.

archive fever is *unlimited*. It recognizes no limits whatsoever, it is absolutely unbounded. In its unboundedness, "archive fever verges on radical evil."[48]

What will happen to archives? Does the extraordinary pace of technological change spell the end of the archive as we know it—or its survival, albeit in other forms? And how will human beings experience this transformation? The answer is that no-one knows. And yet, says Derrida, we should be vigilant and keep our eyes open to that which is to come (*à venir*) and take responsibility for the extraordinary changes that are under way in archival technology. And it is at least possible to ask, as Derrida does, whether the structure of what Freud called the human psychical apparatus, the site of our reading and interpretation of the world, will remain *essentially* unaffected by this archival-technological development. Or whether this structure will be *changed* or, indeed, actually *enhanced*?[49] Even if it turns out not to be affected, Derrida argues, its *content*, the printed content, its *imprint*, its meaning-creating experience, will be affected: "what is no longer archived in the same way is no longer lived in the same way. Archivable meaning is also and in advance codetermined by the structure that archives."[50] Derrida illustrates this thesis by referring to the "geo-techno-logical shock" that would have made the entire archive of nineteenth-century psychoanalysis completely unrecognizable if Freud and his colleagues, instead of writing "thousands of letters by hand," had had access to modern aids to writing such as "credit cards, portable tape recorders, computers, printers, faxes, televisions, teleconferences, and above all E-mail."[51] He notes briefly that psychoanalysis (like a great many other things) would not be what it is today if, for example, email had existed in Freud's day.

If archival technology therefore determines archivable meaning *in advance*, the question of the archive, what an archive is, is a question not of the past but of the future. For this reason, Derrida does not refer to the "concept" of the archive, just as he does not refer to the concept of paper or book: "We have no concept, only an *impression*, a series of impressions associated with a word."[52] Instead of a strictly defined concept, he has an impression, a notion (*notion*) of an endless or undecidable process. The reason for this is, of course, that, for Derrida, a "concept" can never be stable, can never be determined in a neutral or object way, based on knowledge about the past, but is always in the process of being determined in the space between two opposing forces, in this case between archive drive and archive fever. The concept of the archive, as with every concept that Derrida deconstructs, is always burdened by an essential indeterminacy or unforeseeableness that opens onto the future. The archive is not merely knowledge and memory; it is, above all, a

[48] Ibid., p. 39.

[49] Ibid., p. 32.

[50] Ibid., p. 37.

[51] Ibid., p. 33. Derrida's conclusion is no less valid today, when almost everything is stored in digital archives.

[52] Ibid., p. 51.

promise and, like all promises, a symbol of the future. As Derrida writes in connection with the question of the archive:

> It is a question of the future, the question of the future itself, the question of a response, of a promise and of a responsibility for tomorrow. The archive: if we want to know what this will have meant, we will only know in the times to come. Perhaps. Not tomorrow but in the times to come, later on or perhaps never. A spectral messianicity is at work in the concept of the archive and ties it, like religion, like history, like science to a very singular experience of the promise.[53]

For Derrida, the question of the archive, this undefinable technological-political process, around which the entire electronic revolution is centred, is a question about the future. We will never be able to know what will become of archives. We will never be able to know what happens in the space between the archive drive, with its irrepressible desire to preserve, and archival fever, with its infinitely destructive desire to destroy. For this very reason, we must remain constantly on our guard and take responsibility for the enormous transformation in archival technology that is continually ongoing. If deconstruction is an ethics, that ethics is necessarily bound up with the technological archive.

6.3 Total Nuclear War

While the electronic revolution is an example of an event that is still ongoing, total nuclear war is an event that has not yet occurred. It is a hypothetical event for which governments in both East and West not so long ago drew up detailed plans and strategies. This took place during the Cold War, that is to say, from the end of the Second World War up to the fall of the Berlin Wall in 1989 and the sudden collapse of the Soviet Union two years later. Fortunately, this ultimate event never came to pass. After the Wall came down, both sides saw fit to enter into dialogue, and eventually they began to disarm.

During this era of nuclear weapons, strategic military thinking was based on a single doctrine, that of "mutual deterrence," which was widely accepted and regarded as key to preventing a catastrophe. Some idea of the potential scale of a nuclear war can be gained from the fact that "only" 5000 Soviet nuclear missiles would be needed in order to hit every large and medium-size city in the United States with fifty missiles, each with at least forty times more explosive force than the bombs dropped on Hiroshima and Nagasaki at the end of the Second World War. In total, each side had as many as 35,000 missiles.[54]

It was in this bleak climate in 1984 that Derrida published an article in the journal *Diacritics* titled "No Apocalypse, Not Now" (based on a lecture at Cornell University), which came to have a major impact, particularly on the debate over

[53] Ibid., p. 60.

[54] Klein, "Knowledge of The Future", p. 173.

nuclear weapons in the United States. 1984 was undoubtedly a sinister date. Ever since the publication of George Orwell's novel *Nineteen Eighty-Four* in the shadow of the Second World War, it had symbolized a dystopian future. And it was second only to 1962, when the Cuban Missile Crisis had taken place, as the most dangerous year of the entire Cold War. The previous year, President Ronald Reagan had dubbed the Soviet Union an "evil empire "and launched what came to be called the "Star Wars program," that is to say, a plan for creating a defensive shield that would protect the United States from a potential nuclear attack.

In his article, Derrida attempts to describe total nuclear war as a hypothetical event, even as he tries to understand what it means for a culture to live in the shadow of such an event. What characterizes total nuclear war, he argues, is that it has not yet occurred, that it has, in fact, never taken place; even so, during this period it was regarded by both superpowers as an event whose eventualities and risks were the subject of meticulous calculations. But what characterizes this event above all is its *virtuality*. It exists only as something that people can talk about—consciously or unconsciously. Just like a purely literary event, total nuclear war can only be represented in the form of a fictional narrative. It has no real referent, only a *fictive* one. It exists only in the form of a fiction or, as Derrida puts it, a *fable*. As he writes:

> You will perhaps find it shocking to find the nuclear issue reduced to a fable. But then I haven't said simply that. I have recalled that a nuclear war is for the time being a fable, that is, something one can only talk about. But who can fail to recognize the massive "reality" of nuclear weaponry and of the terrifying forces of destruction that are being stockpiled and capitalized everywhere, that are coming to constitute the very movement of capitalization?[55]

Total nuclear war may have been a myth, a fiction, and, for some, even a poem. A virtual event, in any case. Nevertheless, this war had real consequences (which in many regards we are more aware of today). Confronted by the possibility of a total nuclear war, a massive armament programme was prepared, planned, and carried out, involving massive costs in the form of resources and capital. It is therefore necessary to differentiate between the reality of the age of nuclear weapons and the fiction of nuclear war: they are two separate concepts. And yet they are not independent of each other. Nuclear war as fiction, as fable, is the basis for the fabulous military preparations, the pointless stockpiling of vast numbers of weapons, the arms race with no definite goal, and the extraordinary haste, resulting from technology and technological inventions, that affected not only the military, diplomacy, and politics but ultimately all humanity. This reality is based on an event that in reality is a *non-event*: "'Reality,' let's say the encompassing institution of the nuclear age, is constructed by the fable, on the basis of an event that has never happened (except in fantasy)."[56] Exactly like a work of literature, total nuclear war exists only in language—only in what is said about it and when it is talked about. And it is only through language, through a final linguistic event, that total nuclear war can finally be unleashed. This war is "a phenomenon whose essential feature is that of being

[55] Derrida, "No Apocalypse, Not Now", p. 23.

[56] Ibid.

fabulously textual, through and through. Nuclear weaponry depends, more than any weaponry in the past, it seems, upon structures of information and communication, structures of language, including non-vocalizable language, structures of codes and graphic decoding."[57] Nuclear war is fabulously textual, not only because we can only talk and write about it but also because the various parts of nuclear weapons systems have been predetermined to work with information and communications structures. As a hypothetical event, nuclear war is now necessarily spoken and written text that is produced and presented by a teletechnological media apparatus.

At this point Derrida moves on to the question of whether there is any fundamental difference between the destruction caused by a nuclear war and that caused by other kinds of war? What exactly is the difference between nuclear war's fabulously textual event and the total extermination of entire peoples that the world has already experienced? In what sense is this an unprecedented, absolutely unforeseeable event? In what sense is it unique? After all, it is not the first time: other wars have also only ever been written and spoken—so long as they have not taken place. With regard to fears about a future war, Derrida poses the following question:

> What might prove that a European in the period following the war of 1870 might not have been more terrified by the "technological" image of the bombings and exterminations of the Second World War (even supposing he had been able to form such an image) than we are by the image we can construct for ourselves of a nuclear war?[58]

There is something in this comparison, particularly if we imagine a limited and "clean" nuclear war. But this becomes an impossibility if we posit a *total* nuclear war. Unlike previous wars, all of which have been preceded by wars of more or less the same type, total nuclear war has no precursors. It has never taken place (rather than starting a nuclear war, the American bombings of Hiroshima and Nagasaki in 1945 ended a classical, conventional war). Total nuclear war has never taken place, which is precisely why we can only talk and write about it. It is a fable of unlimited destruction, a terrifying event that may perhaps happen, that is perhaps to come—from the unforeseen future that Derrida calls *avenir*. This fable is a human "invention" in both senses: it is not real (and, were it to be realized, it would bring down the curtain on human life on earth) and it is also an invention in the sense of unquestionably having its point of departure in new "technical mechanisms." These technical mechanisms are extraordinarily sophisticated: they are "the technologies of delivery, sending, dispatching, of the missile in general, of mission, missive, emission, and transmission, like all *techne*."[59] According to Derrida, these extraordinarily

---

[57] Ibid.

[58] Ibid.

[59] Ibid, p. 24. A missive is a written consignment. Derrida's article, which is subtitled "Full speed ahead, seven missiles, seven missives", is divided into seven sections, each headed "missile, missive", in the form "first missile, first missive", "second missile, second missive", and so forth. Derrida here connects missile and missive precisely in order to underscore that they both represent something that is *sent* from a sender to a recipient. For Derrida, a missive is not the only written consignment; even a missile, including a nuclear missile, can be described as a written consignment so long as the concept of writing is understood in its general sense. In actual fact, all writing,

sophisticated technological mechanisms have allied themselves with the "sophistry" of the military-political authorities, their "psycho-rhetoric," and "the most vulgar psychology."

Investigating the rhetoric of political texts from the Trojan War to the era of nuclear war, Derrida finds that the "rhetorical situation" is practically unchanged. The technology of war has developed beyond recognition, it is true, but the political rhetoric, the mental structures, and the game-theory calculations are essentially the same. As always, politicians and military leaders in the age of nuclear weapons justified the enormous expansion of nuclear weapons capacity in terms of "achieving victory," "avoiding defeat," and, above all, of converting their superior capacity into "diplomatic power." Derrida reminds us that in the diplomatic moment, in the negotiating situation itself, there is only text, that is to say, text in the general sense that he has given this word but also text in the narrowly conventional sense.

Among all of the kinds of texts that exist, Derrida then chooses a subset that he regards as especially important in the age of nuclear weapons and that he calls *literature*. When he investigates the conditions of possibility for this literature, he finds two defining conditions. The first relates to the construction of an objective archive of every body of oral tradition. This involves gathering, storing, and preserving texts in a large number of institutional archives. As a result, these literary texts attain an objective status that spoken texts never can have. The second condition relates to "the development of a positive law implying authors' rights, the identification of the signatory, of the corpus, names, titles, the distinction between the original and the copy, the original and the plagiarized version, and so forth."[60] In other words, a legal archive comprising consecrated authors' works in which author's name and work are inseparably intertwined. For Derrida, these two conditions are absolutely necessary for the existence of a literature that is, in principle, timeless. Yet these two conditions also point to an interesting relation between literature and nuclear war:

> Now what allows us perhaps to think the uniqueness of nuclear war, its being-for-the-first-time-and-perhaps-for-the-last-time, its absolute inventiveness, what it prompts us to think … is obviously the possibility of an irreversible destruction, leaving no traces, of the juridico-literary archive—that is, total destruction of the basis of literature.[61]

Derrida here entertains the hypothetical possibility of the total eradication of all traces. All that remains is a single, indelible trace, "the trace of what is entirely other, "*trace du tout autre*."[62] This event takes place for the first time and therefore cannot be compared with previous catastrophes such as the burning of libraries, not

all text, and thus all *technē* is ultimately something that is sent, that allows itself to be transmitted through time and space. This does not mean that missiles, like writing, can be reduced to something feeble or inoffensive (as some people naively regard books). Language and writing always carry an explosive force, "the power of a death machine" (p. 29).

[60] Ibid., p. 26.

[61] Ibid.

[62] Ibid., p. 28.

even that of Alexandria (which actually gave rise to a host of literary texts). Unlike other events, this event is impossible to imagine since the first time it takes place is also the last. This is precisely why this event represents the entirely other, the impossible. And, Derrida argues, we must face "the necessity and the impossibility of thinking the event."[63]

Next, Derrida connects this hypothesis about the entirely other, the impossible, the total destruction of the archive, to deconstruction: "The hypothesis of this total destruction watches over deconstruction, it guides its footsteps." In light of this hypothesis, so to speak, it becomes possible to discern the characteristic structures and historicity of that which is to be deconstructed. This is why "deconstruction, at least what is being advanced today in its name, belongs to the nuclear age."[64] This is why the *experience* of the entirely other, the impossible—in this case, the total destruction of the archive—is the very definition of deconstruction. In other words: the hypothesis of the archive's total destruction *governs* deconstruction's ethical engagement.

But why is deconstruction's ethical engagement connected to the destruction of the archive in the first place, rather than the tragic loss of all human life and the annihilation of a habitable environment? Perhaps, despite everything, there will be survivals and perhaps a habitable environment can gradually be reconstructed. But what will be lost without trace and can never be reconstructed is the archive. For Derrida, the archive is the very symbol of the nuclear age's vulnerability because memory itself, the social memory that makes it possible to mourn for the dead, can be destroyed. As he writes:

> The burden of every [individual] death can be assumed symbolically by a culture and a social memory (that is even their essential function and their justification, their *raison d'être*). Culture and memory limit the "reality" of individual death to this extent, they soften or deaden it in the realm of the "symbolic."[65]

Social institutions for commemorating the dead may be the earliest and most fundamental hallmark of a civilization. After a total nuclear war, however, an archive that has been destroyed can never be reconstituted. Even if there were a future after such destruction, it would not be possible to commemorate the past. Total nuclear war will destroy the future itself—that future space in which we normally assume it will be possible for an event to be preserved, mourned, or celebrated. What would be lost if the archive were destroyed, if there were no institutional memory able to commemorate the dead? Total annihilation is not an apocalypse: it is not an apocalyptic revelation, not an unveiling of the truth of being, for the simple reason that there is no memory of the past. An event comes from the future, says Derrida. But total nuclear war is the name of a possible event without future, an event that is not an event.

---

[63] Ibid., p. 30.

[64] Ibid., p. 27.

[65] Ibid., p. 28.

6.4 Global Terrorism

A third event in which Derrida took particular interest was the terrorist attacks on the United States of 11 September 2001. In their book *Philosophy in a Time of Terror* (2003), the material for which was gathered in New York just a few weeks after the attacks of 11 September, Giovanna Borradori interviewed Derrida about the importance of arguably the most discussed and written-about terrorist acts of all time: four American commercial airliners were hijacked, two of which were flown into the twin towers of the World Trade Center in Manhattan, causing them to collapse, the third was crashed into the Pentagon building in Washington, D.C., and the fourth crashed in a field in Pennsylvania. Almost four thousand people, including the nineteen hijackers, lost their lives, most of them in the Twin Towers.

There are, I contend, a number of similarities between Derrida's analysis of this event and his deconstruction of nuclear war. While it may be easier to think of nuclear war as a fable about the future than to think of "11 September" in the same terms, they nonetheless have a comparable relation to an unforeseeable future, to a non-apocalyptic, future catastrophe. For Derrida, it is precisely this relation to the future that lends 11 September such traumatic force. Although it involves an event that has already taken place, it nonetheless resembles nuclear war in conjuring up, consciously or unconsciously, the possibility of future terrorism that is at least as terrifying as the possibility of total nuclear war.

Even though Derrida sees terrorism as directed at the future, the very word terrorism has an older provenance. It can be traced back to the last phase of the French Revolution, when Robespierre's Reign of Terror degenerated into mass executions and purges of civilians. Robespierre was the first (but absolutely not the last) to practise state terrorism; conversely, all terrorists see themselves as heroic freedom fighters whose attacks are a response to what they see as terrorism in the service of an oppressive state. "Every terrorist in the world," Derrida notes, claims to be responding in self-defense to a prior act of terrorism on the part of the state, one that simply went by other names and covered itself with all sorts of more or less credible justifications."[66]

It is hard to ignore the existence of state terrorism as a form of terrorism, even if accusations of state terrorism are sometimes simplistic or simply wrong. But state terrorism is not the only force that must be recognized as a form of terrorism. For instance, asks Derrida, when hundreds of millions of people are left to die of famine, AIDS, lack of medicine, and so forth—should these not also be considered as part of a conscious and deliberate strategy of terrorism? Indeed, does this strategy even need to be conscious or deliberate? Are there situations when terrorism is waged within a system without any single individual being aware of or assuming responsibility for it?

For Derrida, the word terrorism is far too narrowly defined in the discourse that dominates in the public sphere. This narrow definition is principally disseminated

[66] Derrida, *Philosophy in a Time of Terror*, p. 103.

via the techno-economic power of the media, particularly television. A prime example of this narrow definition is 11 September itself. What would 11 September be without television? What would 11 September be without television's continual repetition of the terrorist attacks? According to television, the perpetrators of 11 September were the very embodiment of the concept of terrorism. Nor should we forget that, for different reasons, the terrorist perpetrators of 11 September and those who wished to declare "war on terror" in the name of its victims were both interested in giving the event maximal media coverage.[67]

What, then, should be our attitude towards terrorism? Rather than opting for a strategy of declaring war on terrorism, Derrida chooses what he sees as a more ethically responsible strategy, namely of *deconstructing the concept of terrorism.* His reasoning is that the prevailing use of the concept of terrorism (as a more or less self-evident concept) unintentionally plays into the terrorists' hands. Deconstructing terrorism involves showing that the distinctions we usually associate with the word terrorism are problematic. For example, the concepts of war and terrorism are not mutually exclusive: war has always been contaminated by terrorism in the form of terror used against civilian populations. And, Derrida argues, modern techno-science is blurring the distinction between war and terrorism.[68] Nor is it possible to draw a sharp line between different kinds of terrorism, for example, between national and international, local and global terrorisms. Because we cannot agree on a simple definition of terrorism, we clearly have to accept that the concept does not have a fixed meaning, agenda, or political content. Terrorism is "absolutely unforeseeable and incalculable."[69]

After making this general observation, Derrida proceeds to analyse the 11 September event. What does it mean to refer to an event by a *date*? Why do we call what happened "11 September"? This date is repeated continually, as if it were something utterly unique, something that defies all attempts at generalization. But, he argues, it is not only we who are repeating it to ourselves, like a kind of inner mantra or ritual incantation; it is also repeated endlessly by the media, helped by a vast techno-political machine, the same machine that originally dubbed the terrorist attacks "11 September." The reason for this media repetition, according to Derrida, is that we want to convey the impression that a *unique* and *historically unprecedented event* has taken place. By connecting an event with a date, it automatically becomes historically significant: it becomes a *singular event.*[70]

Fundamentally, then, it is the media and the global communication networks that *create the impression* of a singular event by continually saturating the world with

---

[67] Ibid., p. 108.

[68] Propaganda via radio and television has placed an essential role in "declaring" war ever since the First World War. It has gone hand in hand with bombing raids that, like resistance movements and their use of coercion, were unable to differentiate between civilians and soldiers. Ever since the First World War, it has thus been impossible to make a clear distinction between war and terrorism (Derrida, *Philosophy in a Time of Terror*, p. 109).

[69] Ibid., p. 98.

[70] Ibid., p. 86.

images and reports about 11 September. Our impressions are mainly produced by a gigantic information apparatus that has been political, technological, and economic from its inception. Derrida confesses to being surprised by the naivety with which this apparatus exacerbated the trauma of the general public's experience of 11 September.[71]

Sadly, however, it takes more than the deaths in a few seconds of almost four thousand people, mostly civilians, as a result of modern technology, for something to count as a singular event. There are many examples, not only from the two world wars, of mass murders being committed during a short space of time that have not been regarded, analysed, experienced, or presented as singular events. None of these events have given the impression of being unimaginable catastrophes, mostly because they received patchy or non-existent media coverage.

We must always distinguish between two kinds of impression. On the one hand, a media-produced impression that makes us *believe* that 11 September is a singular event. On the other, a genuine sympathy with the victims and a deep outrage at the deaths; a boundless and unreserved grief at the loss of human lives and the vulnerability of the system that is meant to protect against terrorism. This system is embodied in the world's only superpower, the United States, which is both the location where the attacks occurred and the enforcement agent of the current world order. In its role as the leading techno-scientific, capitalist, and military power in the world, the United States (for now, at least) is itself the symbol of the current world order. According to Derrida, this world order is built on the decisiveness, reliability, and credibility of American power. To expose the vulnerability of this superpower is also to show up the vulnerability of the world order.

Derrida interprets this vulnerability as an *auto-immune crisis* within the system that should have prevented it. *Auto-immunity* is a concept that he has borrowed from the life sciences, where it refers to the harmful production of antibodies by the immune system against its own organism. In Derrida's thinking, this becomes a mechanism by which a state "works to destroy its own protection, to immunize itself against its "own" immunity."[72] The state destroys itself precisely by protecting itself; its protection is both cure and toxin. The Greek word for this is *pharmakon*, Plato's term for this auto-immune logic. In our era, it shows itself "in the inevitable perversion of technoscientific advances (mastery over living beings, aviation, new informational teletechnologies, e-mail, the Internet, mobile phones, and so on) into weapons of mass destruction, into 'terrorism' of all kinds."[73]

Derrida distinguishes *three phases* in the auto-immune crisis of the West.[74] The first phase is the Cold War, which was fought in the minds of the two superpowers rather than on the ground or in the air. If we regard 11 September as an extension of the Cold War, it is easy to understand how this auto-immune mechanism functions:

---

[71] Ibid., p. xiii.

[72] Ibid., p. 94.

[73] Ibid., p. 124.

[74] Ibid., p. 94.

the hijackers who attacked the United States had been trained by the United States to resist the Soviet occupation of Afghanistan. American weapons and intelligence were placed at the disposal of Afghan resistance fighters, which effectively contributed to the growth of Islamic fundamentalism and, eventually, the formation of the al-Qaida terrorist network.

11 September is thus a symptom of an auto-immune crisis, according to which the symbol of the current world order, the United States, is the victim of aggression that comes *from without*. At the same time, this aggression also comes *from within*, from forces clearly lacking their own resources but who used ingenuity and high-tech know-how to smuggle American weapons onto American aircraft in American cities. 11 September can thus be thought of as the Cold War's grand finale, as an implosion of inner convulsions and contradictions.

The second phase of the auto-immune crisis is, according to Derrida, "worse than the Cold War."[75] Whereas the Cold War was characterized by a balance of terror between two superpowers (the United States and the Soviet Union), inseparably joined in a shared understanding of the pointlessness of nuclear war, it is impossible to achieve such a balance with terrorism because the threat no longer comes from a state but from "anonymous forces that are absolutely unforeseeable and incalculable."[76] Yet we now know that uncontrolled proliferation of nuclear weapons and relative availability of bacteriological and chemical weapons are the reality in which global terrorism operates. This reality seems particularly terrifying in light of the unacceptable strategy of terrorists. What strikes Derrida as unacceptable about the terrorists' strategy is not only its cruelty, disregard for human life, lack of respect for laws, for women, or its appropriation of the worst aspects of techno-capitalist modernity for the purposes of religious fanaticism. No, he says, the most unacceptable is that their thinking *is not open to the future*. It has no future (at least, not in *this world*).[77] Their only stated ideology is "the rejection of modernity and secularization."[78]

Naturally, it may be asked exactly how Derrida can say that the terrorists *on the one hand* have no future and *on the other* that they see themselves as heroic freedom fighters. Is it so obvious that the terrorists, as heroic freedom fighters, do not want to attain particular goals, such as putting an end to a globalization that they see as entirely determined by the West? Islamic fundamentalism may come in different guises but don't all terrorists look forward to a world defined by theocracy, Sharia law, and religious fanaticism? Unquestionably. Derrida condemns all of this, naturally. What he means is not that they cannot see a future—which they do—but that their thinking is not open to the other, to the secularization that the world has so laboriously developed in the form of politics, democracy, international law,

---

[75] Ibid., p. 94.

[76] Ibid., p. 98.

[77] Ibid., p. 113.

[78] Ibid., p. xi.

international institutions, and so forth. Not the slightest part of this seems to figure in the way terrorists think.[79]

For Derrida, nothing good can be hoped for from the terrorists. Rather, we must expect an absolute evil, whose threatening shadow is growing in every direction. Since this absolute threat targets the powers that guarantee the current world order, the United States and its allies, what is at stake in this auto-immune terror is "nothing less than the existence of the world, of the worldwide itself."[80]

The fact that trauma is today primarily targeted not at us or at the past but at the *future*, at that which is to come, at *l'avenir*, makes such terrorism psychologically worse than the Cold War. A weapon injures and leaves an open wound that can be mourned forever, but *these* weapons are terrifying because they come from an absolutely unforeseeable and unpredictable future. Imagine, says Derrida, that you receive the following message: what has happened—two skyscrapers have collapsed and thousands of people have died in just a few seconds—is of course horrific, a terrible crime, an incomparable calamity, but it is all over now; nothing this horrific, or even worse, will ever happen again. How would we react? Derrida assumes that the work of mourning would in this case be completed within a relatively short space of time, after which life would fairly quickly go back to normal. People would have mourned for a while and then moved on (as has often happened, and relatively easily, in the wake of catastrophes in other parts of the world, far from Europe and the United States). But this is not what happened at all. What makes 11 September so traumatic is not that the aggression has been committed and concluded but that trauma is being produced by *the future*, by the threat that *the worst may be yet to come*. It is this trauma, this dread of future attacks that makes 11 September *look like* a singular event.[81]

Derrida calls the third and final phase of this auto-immune crisis "the vicious circle of repression."[82] It is by far the most dangerous because it describes how the United States and its allies, in declaring war on terrorism, have created a kind of war against themselves. According to Derrida, it is indefensible to try to eliminate trauma by means of war and oppression. We can never forget the implacable logic of autoimmunity: that suppression in both its psychoanalytic and political senses—whether through the police, the military, or the economy—sooner or later produces, reproduces, and regenerates the very thing it seeks to eliminate.

In order to break the vicious circle of war and oppression, our need for political leadership and media that behaves in an ethically responsible fashion is going to increase in the future, particularly in light of what Derrida fears will be the future of terrorism: virtual attacks. In his view, techo-science has changed the relationship between terror, terrorism, and territory—three words with a common root in the

---

[79] Ibid., p. 133–4.

[80] Ibid., p. 99.

[81] Ibid., p. 96–7.

[82] Ibid., p. 100.

Latin word *terra*. Future threats are not only territorially indeterminate, their violence is completely different and more terrifying because it is silent and invisible:

> when compared to the possibilities for destruction and chaotic disorder that are in reserve, for the future, in the computerized networks of the world, "September 11" is still part of the archaic theater of violence. … One will be able to do even worse tomorrow, invisibly, in silence, more quickly and without any bloodshed, by attacking the computer and informational networks on which the entire life (social, economic, military, and so on) of a "great nation," of the greatest power on earth, depends. … Nanotechnologies of all sorts are so much more powerful and invisible, uncontrollable, capable of creeping in everywhere. They are the micrological rivals of microbes and bacteria. Yet our unconscious is already aware of this; it already knows it, and that's what's scary.[83]

There are worse scenarios than two passenger planes flying into skyscrapers and causing them to collapse. Terrorist attacks no longer need planes, bombs, or kamikaze pilots: it is enough to infiltrate a strategically important computer system and infect it with a virus or some other damaging element to paralyze the economic, military, and political life of an entire country or continent. Future attacks—with chemical, biological, or nuclear weapons or by completely paralyzing the digital information system—will be able to be launched from virtually anywhere in the world, at a trifling cost and with minimal effort—silently, invisibly, and unpredictably.

Confronted by the deadly threat represented by global terrorism and total nuclear war, Derrida seeks a global answer, an *ethical responsibility*, which involves a transition from classical international law, with its roots in nineteenth-century legislation of nation states, to a new, *cosmopolitan* world order in which the key political actors are international institutions and continental alliances. He is well aware that such a transition is infinitely complicated, even utopian bordering on impossible. Even so, he argues that we must dare to believe in the possibility of this impossibility, and allow this possibility to be the ethical imperative that guides us in all our decision-making.

The cosmopolitan world order that Derrida seeks, beyond the nation state and beyond all ethnic and religious affiliations, is being produced by the extraordinarily rapid technological transformation ushered in by the electronic revolution. This accelerating development of communication and information technology, what Derrida calls the teletechnological apparatus, is the very motor of the process of political transformation. Indeed, Derrida claims that teletechnologies *produce* the deconstruction of the political: "What the accelerated development of teletechnologies, of cyberspace, of the new topology of "the virtual" is producing is a practical *deconstruction* of the traditional and dominant concepts of the state and citizen."[84]

However, teletechnology's practical deconstruction of the nation state does not mean that the cosmopolitan ideal, this Judeo-Christian ideal that goes back to the Stoics, St. Paul, and Kant, cannot or should not be criticized. The cosmopolitan ideal and its goal of universality, Derrida argues, must be adapted to our era by calling into question the way in which this ideal has been determined by the

[83] Ibid., p. 101–102.

[84] Derrida & Stiegler, *Échographies de la télévision*, p. 45.

"ontotheological, philosophical, and religious discourses in which this cosmopolitical ideal was formulated."[85] Although he supports the idea of world citizenship and cosmopolitan rights, it is our ethical duty, he argues, to move beyond the cosmopolitan ideal itself, beyond the idea of world citizenship, in fact, even "beyond the 'political' as it has been commonly defined."[86] This "beyond the 'political,'" beyond democracy here-and-now, is what Derrida calls *la démocratie à venir*.

6.5 Summary

In this chapter, I have chosen to show that the ethical dimension of Derrida's thought is closely bound up with technology. I have also demonstrated what this connection consists of and how it should be understood, and I have illustrated it using three world-historical events: the electronic revolution; total nuclear war; and global terrorism.

I began by noting that there is now a strong consensus among commentators on Derrida that deconstruction *is* an ethics. This means that his entire philosophical orientation is ethical in the sense that he understands the word. For him, ethics is not about a set of ethical rules or prescriptions. It is neither a normative ethic nor a *theory* about ethics but an ethics of ethics, the very condition of possibility for all ethics. It is an originary ethical experience of the other, the entirely other—*le tout autre*. In this deeper sense, ethics involves a command, an imperative, a challenge to be open to the entirely other.

Such openness to the entirely other is always an aporetical experience because it presupposes a mode of thinking that does not reduce the other's alterity to pure identity. This aporetical experience is what Derrida explicitly defines as *deconstruction*. The fact that deconstruction is an aporetical experience means that it has a temporal structure, that it is a mode of thinking in time. But even if it operates in time, empirical time, it is in fact engaged with the future, with the absolute and structural future that Derrida calls the to-come—*l'avenir*. Deconstruction is continually open to this future, to that which comes to us from without, suddenly, from a place that we can neither know nor predict.

That which comes to us in this way is an *event*. The event is not something that happens in some very general way, but the arrival of the completely other, which we can never foresee, which it is *impossible* for us to imagine. This does not mean that there is anything mysterious or supernatural about the arrival of such an im-possible event. Even if it comes from the future—*l'avenir*—it must, like all things that happen, take place in time. Since time must be inscribed as traces in space, our experience of the event is dependent upon the *material substrates and the technologies* that are associated with the event. These traces, these representations of technology,

[85] Derrida, *Philosophy in a Time of Terror*, p. 130.

[86] Ibid.

constitute all experience through movement in *différance*. In other words, every event is essentially *technically mediated.*

Although the event can be experienced *live*, that is to say, in real time, it is increasingly mediated by media. Now more than ever, says Derrida, the event is produced and presented by a gigantic teletechnological media apparatus. This apparatus creates the event, it produces it, interprets it, and perhaps even reinterprets it in accordance with its own interests: the event becomes an *artifact.*

Derrida argues that it is necessary to deconstruct this artifactual production. It is necessary because deconstruction involves a demand that someone take responsibility for the underlying event, the real event as it actually happened. Moreover, it is necessary, indeed, a duty, to focus deconstruction on *future* technological events, as these will blur the borders of nation states and thereby challenge the territorially defined concept of democracy. We must therefore try to imagine democracy in a new way, beyond the boundaries of the political. We must imagine the political beyond politics and the democratic beyond democracy. We have been given this ethical imperative—concretely, urgently, daily—by technology itself. If deconstruction is an ethics, then this ethics is necessarily bound up with technology.

The three events of world-historical importance that I have discussed in this chapter illustrate precisely this necessary connection between ethics and technology. The first example—the electronic revolution—involves a transition from a centuries-old paper culture to a global electronic culture. From the beginning, Derrida was deeply engaged with this "geo-techno-logical shock": he deconstructed the concept of paper constantly and he wondered anxiously what happens when a world of paper is replaced by apparatuses or machines for electronic inscription. His deconstructive analysis of the book starts from the question of what will happen if and when the *future* book comes to us. The same engagement is again present in his deconstruction of the *archive*, the central focus for everything in both paper culture and electronic culture. All in all, deconstruction is deeply engaged with the technology that comes from the future and that is replacing "paper": the future book, the future archive, and, indeed, the transformative power of the entire electronic revolution.

The other example is total nuclear war. This terrifying event is a hypothetical event, since it has never actually taken place. If and when it does take place, it will be for both the first and the last time. For this reason, it represents the completely other, the impossible that we cannot imagine. We must, Derrida argues, take responsibility for the possibility of this impossibility—the irrevocable and traceless *destruction of the archive*, that is to say, the total destruction of collective memory, of history. Deconstruction's ethical engagement is thus grounded on the hypothesis of the archive's total destruction. After a total nuclear war, the destroyed archive can never be reconstructed. This event is completely unique, not because of the tragic loss of all human life and the annihilation of a habitable environment (there will perhaps be survivals and perhaps a habitable environment can be reconstructed) but because there would no longer be any memory of the past. Total nuclear war will destroy the future itself—the future place from which we normally assume that an event can be preserved. Total nuclear war is therefore the name for a possible event without a future, an event that is not an event.

The third example concerns the terrorist attacks against the United States on 11 September 2001. As in the case of total nuclear war, "11 September" has a relationship to a possible future catastrophe. It is precisely this relationship to the future that lends 11 September its traumatic power. Even though it involves an event that has already happened, it nonetheless resembles nuclear war in that it conjures up the prospect of a *future* terrorism that is every bit as terrifying as the prospect of total nuclear war. One of the aims of this future terrorism is to show the vulnerability of the world order that was intended to protect us from terrorism.

Derrida describes this vulnerability as an *auto-immune crisis* within the very system that was supposed to prevent it. He differentiates *three phases* of this auto-immune crisis. The first is the Cold War, which was characterized by a balance of terror between the United States and the Soviet Union and that came to an end with the attacks of 11 September; the second is global terrorism, which represents not a state but anonymous forces that are absolutely unforeseeable and unpredictable. The third and final phase is by far the most dangerous because it describes how the United States and its allies, by declaring war on terrorism, are creating a kind of war against themselves. It is impossible, Derrida argues, to escape the implacable logic of auto-immunity: that war and oppression sooner or later produce the very thing they are trying to eliminate.

In Derrida's view, the deadly threat represented by global terrorism and total nuclear war calls for a global response in the form of a new, cosmopolitan world order in which the key actors are international organizations and regional alliances. This does not mean that this cosmopolitan idea, with its roots in antiquity, should not be criticized; on the contrary, it is our ethical obligation to go beyond the cosmopolitan ideal itself, indeed, to go beyond the political, a topic that will be examined in greater detail in the next chapter.

References

Critchley, S. (1992). *The ethics of deconstruction: Derrida and Levinas*. Blackwell.
Derrida, J. (2003a). *Voyous: Deux essais sur la raison*. Galilée.
Derrida, J. (2003b). In G. Borradori (Ed.), *Philosophy in a time of terror*. The University of Chicago Press.
Derrida, J. (2002). Ethics and politics today. In E. Rottenberg (Ed.), *Negotiations: Inventions and interviews, 1971–2001*. Stanford University Press.
Derrida, J. (2001). *Papier Machine*. Galilée.
Derrida, J. (1999). *Donner la mort*. Galilée.
Derrida, J. (1996). *Apories*. Galilée.
Derrida, J. (1995a). *Mal d'archive: Une impression freudienne*. Galilée
Derrida, J. (1995b). *Archive fever: A Freudian impression*. (trans. E. Prenowitz). The University of Chicago Press.
Derrida, J. (1993). *Aporias*. (trans. T. Dutoit). Stanford University Press.
Derrida, J. (1987). *Psyché, invention de l'autre*. Galilée.

Derrida, J. (1984). No apocalypse. Not now (full speed ahead, seven missiles, seven missives). *Diacritics, 14*(2), 20–31.
Derrida, J. (1976). *Of grammatology*. (trans. G. C. Spivak). The John Hopkins University Press.
Derrida, J. (1967a). Violence et métaphysique. In *L'écriture et la difference* (pp. 117–228).
Derrida, J. (1967b). *De la grammatologie*. Minuit.
Derrida, J., & Ferraris, M. (2001). *A taste for the secret*. (trans. G. Donis). Polity Press.
Derrida, J., & Stiegler, B. (2002). *Echographies of television: Filmed interviews*. (trans. J. Bajorek). Polity Press.
Derrida, J., & Stiegler, B. (1996). *Échographies de la télévision*. Galilée.
Klein, R. (2008). Knowledge of the future: Future fables. *Diacritics, 38*(1-2), 173–179.
Rosenfeld, M. (2006). Derrida's ethical turn and America: Looking back from the crossroads of global terrorism and the enlightenment. *Cardozo Law Review, 27*, 815–845.

Chapter 7
Politics and Technology

7.1 Introduction

At the start of "Violence et métaphysique," his brilliant essay on Levinas and his critique of phenomenology, Derrida describes two separate historical patterns of thought: on the one hand, the Greek *logos*, whose conceptual universality encloses the realm within which philosophy is possible; on the other, the Judaic wisdom that seeks to escape the clutches of *logos*.[1] The essay remains evasive and ambivalent about these two patterns of thought: the Greek—rational, orderly, universal, and forming the basis for laws and politics; and the Judaic—ethical and emphasizing experience of the Other's alterity. Although the essay avoids taking a stance on Judaic ethics and Athenian politics, it seems for that very reason even more necessary to try to understand the things that unite and divide them. A connection is necessary (*il faut ce rapport*), Derrida argues, even as he demands that we make space for them both; these are *separated* domains. There must be a gap—or rather, a chasm—between ethics and politics that makes it impossible to move safely from one to the other.[2]

Surprisingly, for Derrida, this chasm is not something negative but, on the contrary, both ethically and politically welcome. In true deconstructive style, he defines ethics as "nonnegotiable" and politics as the art of negotiating about the "nonnegotiable."[3] Unlike ethics, which is unconditional, politics is conditional and must respect the prevailing power relations, current laws, possible alternative actions, and so forth, something that requires comprehensive analysis and strategic thinking.[4] The chasm between ethics and politics makes it possible for Derrida to embrace

---

[1] Derrida, "Violence et métaphysique", p. 122.

[2] Derrida, *Adieu*, p. 196.

[3] Derrida, "Ethics and Politics Today", p. 304.

[4] Ibid., p. 301.

B. Sjöstrand, *Derrida and Technology: Life, Politics, and Religion*, https://doi.org/10.1007/978-3-030-83407-4_7

ethical responsibility while *also* prising open a political sphere that is despotically closed. Politics thereby becomes an area of risk and danger, requiring not only political ingenuity but a capacity for *taking decisions*. According to Derrida, only decisions of this kind, with no guarantees whatsoever, can be taken in full freedom.

Politics cannot be derived from ethics because any such derivation would circumscribe the freedom of political decision-making. In politics, decisions must be taken under conditions of greatest possible uncertainty. Politics must always be open to what *may happen*, may perhaps come to pass, the to-come. And, he continues, nothing does greater *justice* to the to-come than this "perhaps."[5] Thinking in terms of "maybe" marks a break with a mode of thinking that proceeds from the assumption that political decisions are ultimately no more than applying certain self-evident programmes or rules of action. This kind of thinking would deprive political decisions of their character as freely-taken decisions. For Derrida, nothing could be more irresponsible and totalitarian than to exclude evil from politics in advance: "Without the possibility of radical evil, of perjury, and of absolute crime, there is no responsibility, no freedom, no decision."[6]

Nor can politics be despotic. If it were, it could be derived from the notion of a voluntaristic and sovereign will, in which the possibility of taking decisions presupposes the existence of a sovereign subject. This kind of Schmittian "politics" is a politics without ethics.[7]

If politics cannot be derived from ethics and is not despotic, we can follow Derrida in taking a decisive step, namely, separating ethical *form* from political *content*. On the one hand, we can embrace the *formal* notion of responsibility for the other; on the other, we can also take concrete decisions about political *content*. Taking ethical-political responsibility would then involve, in every specific situation, filling the *form* of ethics with political *content*. In so doing, we can reconcile ethical responsibility with political decision-making.[8]

Because Derrida regards ethics and politics as *separate* domains, I have chosen, for pedagogical reasons, to divide my analysis into two chapters, one dealing with the relation between ethics and technology (the previous chapter), and one investigating the relation between politics and technology (the present chapter). The question posed in this chapter is whether modern technology is only one of several factors that define contemporary politics or whether this politics is *essentially* bound up with *this* technology in particular.

I begin with a survey of Derrida's politics, his political orientation, and deconstruction's political ambitions. I then discuss his main thesis, namely, that technological development, by erasing social, cultural, and national boundaries, is producing a practical deconstruction of the concept of the political, forcing us to see the political *beyond* politics and beyond the political sovereignty of nation states.

---

[5] Derrida, *Politiques de l'amitié*, p. 46.

[6] Ibid., p. 247.

[7] Ibid., pp. 86–87.

[8] Derrida, *Adieu*, pp. 198–9.

This radical transformation of the political is also leading to the emergence of religious fundamentalism, national archaism, and ethnic separatism, which must therefore be understood as reactions to the erosion of traditional boundaries by accelerating technological progress. But technology is not only erasing these boundaries, it is also erasing the boundary between the political and the private. Recent revelations about the electronic eavesdropping and storage of almost all the whole world's telephone and computer traffic by authorities in the United States and elsewhere clearly illustrates how the political sphere is experiencing a growing need to covertly monitor, manipulate, and control everything that takes place in the private sphere and is stored in private archives. Yet this effort to monitor and control is not limited to private archives but is also directed towards central political or public archives. In a separate section, I therefore highlight Derrida's view that *the archive*, the archive as such, is always already political, always already integrated into a violent political power game. For Derrida, the question of an "archival politics" or a "politics of the archive" is not just a key political question, it is *the* key political question. For this reason, I conclude by examining Stiegler's and Derrida's discussion of the need for what Stiegler calls a *politics of memory*, a political will—if necessary via state intervention—to counter the "industrialization" of memory, which, Stiegler argues, is the result of the hegemony of global techno-capitalism.

7.2 Derrida on Politics

There has been considerable speculation about a so-called political "turn" in Derrida's later writings. In his 2003 book *Voyous*, however, he himself denied such speculations once and for all: "there never was in the 1980s or 1990s, as has sometimes been claimed, a '*political turn*' or '*ethical turn*' in 'deconstruction,' at least not as I experience it. The thinking of the political has always been a thinking of *différance* and the thinking of *différance* always a thinking of the political, of the contour and limits of the political."[9] While it may be true that Derrida only at a late stage began to write *explicitly* about politics, his deconstruction has never confined itself to philosophical analyses of language and text. Ethical-political questions were present in his texts from the very beginning, albeit in an oblique and evasive fashion. In fact, he had good reasons for caution. During the highly politicized period in French intellectual life before and after 1968, he consciously refrained from taking part in the debates over Marxism that dominated political philosophy in France, partly in order to avoid questions about how his work related to dialectical materialism and Marx's philosophy of history. Objections to Marx were viewed as a sign of anti-communism, which was taken extremely seriously in these quarters.[10] Derrida describes a period of "war by other means," a period of avoidance, isolation, and

[9] Derrida, *Voyous*, p. 39.

[10] Derrida, "Politics and Friendship", p. 156.

silence: no-one is quoted, no-one is mentioned by name, people do not communicate. Derrida's silence was also due to his philosophical reservations about the way in which these debates were conducted. He knew that they formed part of an attempt to reform the French Communist Party, and he did not want to disrupt these efforts by bringing to bear on Marx and Marxism the kind of deconstructive questions that he readily engaged with in relation to other philosophers during this period.[11]

However, in the early 1980s, after the collapse of revolutionary Marxism as a force to be reckoned with in world politics, Derrida began writing more about politics. He took up a number of political themes and adopted positions on several political questions of the day. He wrote about deconstruction's political ambition, about law and justice, cosmopolitanism and forgiveness; about political friendship and democracy and a host of other key concepts in modern political thought. He took up the cause of jailed intellectuals and authors around the world, he demonstrated against apartheid, and he adopted strong positions on urgent political questions such as the treatment of illegal immigrants, political reconciliation processes, and the abolition of the death penalty.

In the aftermath of the fall of communism in Eastern Europe, when Marxism was generally regarded as a dead political force, Derrida published his essay *Spectres de Marx* (1993), in which he claimed that deconstruction involved a radicalization of *the spirit of Marxism.*[12] This spirit of Marxism should nonetheless be differentiated "at once from Marxism as ontology, philosophical or metaphysical system, as 'dialectical materialism,' from Marxism as historical materialism or method, and from Marxism incorporated in the apparatuses of party, State, or workers' International."[13] Derrida was fully aware both of Marxism's momentous history as a source of inspiration and of the contradiction inherent in the call to remain true to the legacy of a revolutionary tradition. But this did not prevent him from developing in *Specters of Marx* the idea of a new international as both a counterbalance to global capitalism and a guarantor of international peace between democratic states governed by law.

Although Derrida's new international is utopian and overly abstract, he appears as a kind of cosmopolitan, waiting for the technologization of the world to steadily restrict the sovereignty of the nation state, for better or for worse. He welcomes such restrictions, which he believes will facilitate humanitarian interventions, the abolition of the death penalty, the prosecution of crimes against humanity, and the activities of particular international institutions: "We should salute what is heralded today in the reflection on the right of interference or intervention in the name of what is obscurely and sometimes hypocritically called the humanitarian, thereby limiting the sovereignty of the State in certain conditions."[14] Derrida's welcoming of such restrictions of state sovereignty are sometimes balanced by his defence of the sovereign nation state, for example, as regards the struggle against global linguistic

[11] Ibid., p. 158–64.

[12] Derrida, *Specters of Marx*, p. 105.

[13] Ibid., p. 86.

[14] Ibid., p.105 (translation slightly modified).

dominance, against monopolization and standardization, and for diversity and multiculturalism (nationally and internationally).[15]

But this nuanced defence of the nation state should absolutely not be confused with nationalism. On the contrary, Derrida instinctively harbours a mistrust not only of self-righteous nationalism but of every kind of community, indeed, of the very word community (*communauté*). In *Politiques de l'amitié*, he wonders "why the word 'community' … why I have never been able to write it, on my own initiative and in my name."[16] He was also suspicious of all forms of party politics and of fundamental political categories such as class, race, and gender. In *A Taste for The Secret* (2001), he demanded to be left out of the self-absorbed, exclusive group: "Do not consider me 'one of you,' 'don't count me in,' I want to keep my freedom, always, this, for me, is the condition for being singular and other, but also for entering into relation with the singularity and alterity of others."[17]

Although Derrida never tried to hide his loyalty to the political left, it took the form not of narrow group egoism but of an ethical-political responsibility for alterity in all its forms. He also embraced the Enlightenment ideals of liberation and democracy. He believed in progress and the possibility of progress in politics. But "progress" here must be understood in a formal and negative sense. In *Specters of Marx*, he advocates a "radical and interminable … a necessarily indeterminate, abstract, desert-like experience that is confided, exposed, given up to its waiting for the other and for the event."[18] But Derrida's own political convictions should not be confused with deconstruction's political ambitions. As he saw it, there was no clear connection between deconstruction and fixed political ideas or perspectives. This resulted in critics from a range of positions, including Richard Rorty, Jürgen Habermas, and John Searle, accusing deconstruction of being a variety of nihilistic obscurantism or irrationalism, entirely lacking in principles and unable to serve as the basis for political action. They likewise argued that its strong emphasis on *différance* leads to indecisiveness and a refusal to choose sides—and thence to political relativism.

Derrida responded to his critics by claiming that having a good conscience on the issue of how to think and behave is not desirable; on the contrary, we should call into question and challenge what are currently considered self-evident ways of thinking and behaving. To those who criticized deconstruction for being an "inherently conservative" activity, he replied that deconstruction is inherently neither "conservative" or revolutionary but "inherently nothing at all": "The logic of essence … is precisely what all deconstruction has from the start called into question."[19] As a reply to the claim that deconstruction has no principles, he claimed that deconstruction stands for what it does, namely, calling into question the exist-

[15] Derrida, *Négotiations*, p. 67.

[16] Derrida, *Politiques de l'amitié*, p. 338.

[17] Derrida, *A Taste for the Secret*, p. 27.

[18] Derrida, *Specters of Marx*, p. 112.

[19] Derrida, *Limited Inc.*, p. 141.

ing foundations of political thought and action. Since these foundations are also limits that must be superseded if anything new is to emerge, deconstruction can from this perspective be regarded as a revolutionary gesture. And yet if we want to understand what is at stake in Derrida's deconstructive texts as well as his political actions, the binary opposition of conservative-revolutionary is utterly unhelpful. In *Force de loi* (1994), he champions the idea of "a deconstruction that would like, in order to be consistent with itself, not to remain enclosed in purely speculative, theoretical, academic discourses but rather … to aspire to something more consequential, to *change* things and to intervene in an efficient and responsible, though always, of course, very mediated way."[20]

As a political ambition, deconstruction always involves change and innovation. Its ambition is always to question existing political structures in the attempt to create something new, something wholly other. It is political in the sense that it opens up the present to "let the other come." It prises open deadlocked political structures and concepts so as to forge a path to something wholly other. We will now see how technology itself deconstructs the concept of the political and, with it, the concepts of democracy and citizenship, and how it forces us to think "the political beyond politics" or "the democratic beyond democracy."

7.3 The Transforming of Democracy, Politics, and Citizenship by Technology

Let us start by noting that the idea that politics is connected to technology has a long provenance. In fact, as Bernard Stiegler notes, political citizenship has always had a technological aspect.[21] As far back as the Greek *polis*, having a command of the technology of writing—which made it possible to take part in political life—was a condition of citizenship. Yet the technology of antiquity differs radically from the technologization and digitalization of our own era, which are annihilating time and space and ratcheting up the pace of human societies. According to Derrida, this technologization of the world, which is also accelerating, is changing politics, for better and for worse.

At one point in *Échographies de la télévision*—the filmed discussion between Derrida and Bernard Stiegler—the conversation turns to the way in which the concepts of democracy, politics, and citizenship are being transformed by modern teletechnologies (television, telephony, and other *tele*-communications systems for ultra-rapid transfers across great distances). Derrida argues that our current concept of democracy is "governed, controlled, and limited" by the borders of the nation state, its territorial boundaries, and that if politics is based on citizenship as defined

[20] Derrida, *Force of Law*, p. 8.

[21] Derrida & Stiegler, *Échographies de la télévision*, p. 63. Stiegler uses the French word *caractère*, which also means a written character.

in terms of place, territory, or nation, then teletechnologies will erase these territorial boundaries and, at some undetermined point in the future, perhaps even dissolve them entirely. Demands for national sovereignty and claims to citizenship and democratic rights, all of whose historical legitimacy is inscribed within the frame of topographical or spatial boundaries, will perhaps lose their meaning as a result of the impact of modern technology: "What the accelerated development of teletechnologies, of cyberspace, of "the virtual" is producing is a practical *deconstruction* of the traditional and dominant concepts of the state and citizen (and thus of the "political) as they are linked to the actuality of a territory."[22]

Derrida's argument is neither speculative nor abstract. These territorial boundaries, and the political and legal frameworks on which they are founded, seem increasingly problematic the more we use boundary-crossing telecommunications and computer services. As Derrida observes: each and every time we turn on a TV set, make a phone call, or establish an Internet connection, we are actualizing the question "of democracy, of the political, of deterritorialization." The prefix "tele-," which is a hallmark of modern technology, not only bridges vast distances, it "*displaces places*" (*déplace les lieux*), it compresses space, and transforms the stable spatiality of each individual locus into a general problematic of "spacing" (*espacement*).[23]

In the eighth of his ten theses in *Specters of Marx* about the most pressing problems of the new world order, Derrida addresses this very problematic. His thesis relates to ethnic wars, which continue to flare up despite the claim that globalization is leading to a more homogeneous world. He accounts for this contradiction by referring to the movement in *espacement* and *différance* that operates in the two, equally originary, concepts of "archaism" and "dislocation" (*dislocation*). Archaism designates the ancient notion of rootedness, of a community of the people, national sovereignty, fatherland and ties of blood, the powerful feeling of existence connected to home, place, and people. Dislocation is the rootlessness that is the opposite term of archaism. Even if the process of dislocation is currently spreading in an unprecedented fashion, it is

> no less arch-originary, that is, just as "archaic" as the archaism that it has always dislodged. This process is, moreover, the positive condition of the stabilization that it constantly relaunches. All stability in a place being but a stabilization or a sedentarization, it will have to have been necessary that the local *différance*, the spacing of a displacement gives the movement its start. And gives place and gives rise [*donne lieu*]. All national rootedness, for example, is rooted first of all in the memory or the anxiety of a displaced—or displaceable—population. It is not only time that is "*out of joint*," but space in time, *spacing*.[24]

This is unquestionably an acute and important observation. Derrida's point is that there is an anxiety in both rootedness and rootlessness. An anxiety about not being master or mistress of one's own house. To be rooted in one place is to risk

---

[22] Ibid, p. 45.

[23] Ibid., pp. 68, 76–77.

[24] Derrida, *Specters of Marx*, p. 103.

being dislocated, since there can be no dislocation without rootedness. A people risks being dislocated the moment it roots itself in an archaism.

There is thus an anxiety in rootedness, too. The very idea of rootedness—which is central to all archaism, whether religious, national, or ethnic—is thoroughly spectral. There is always a spectre coming from the future, the possibility of the root's death and a life in some dislocated place and time. War can even break out when this local *différance* escalates to the point of collective affect. Borders become the locus of an exorcism of this spectre from the future. *If we can just secure our borders*, runs the familiar war-cry, *we need never again be fearful for our nation and culture.*

Accelerating teletechnology undeniably creates a feeling of not being master or mistress of one's own house (or of one's own psyche, as Freud would say). This feeling lies behind archaism's reactive and often militant desire to protect the sovereignty of the nation state, strengthen ethnic and religious identity, and secure its own borders (even though, paradoxically, the state's own deregulation of the telemarket is erasing those borders). But such reactive measures are everywhere unable to prevent modern teletechnology from driving an increasingly differentiated and accelerated process of dislocation with such force that it is becoming almost impossible to determine the time and location of the technological event.

At the same time as teletechnology is in this way leading to what Derrida regards as a desirable reduction in the nation state's almost absolute sovereignty and a dissolution of the current national borders, the world finds itself confronted by enormous challenges: an unlimited mobility of people, cultures, epidemics, capital, new waves of migration, populations of homeless and stateless refugees, new forms of terrorism, increased trade in arms and drugs, new forms of colonization, and so forth.[25] Derrida is well aware of these challenges and by way of an answer sees what he earlier on described as a new *cosmopolitan* world order, in which international institutions and continental alliances become the overarching political actors. While his proposals are necessarily sketchy and verging on utopian, they undeniably represent the concrete starting point for more developed solutions to the problems created by teletechnology.

Returning to Derrida's and Stiegler's conversation in *Échographies de la télévision*, it may be asked whether modern teletechnology, which thus reduces the nation state's sovereignty and erases its borders, is merely *one* of several characteristic features of politics today, or whether the political—as Derrida claims—is *essentially* bound up with this technology. For Stiegler, Derrida's position means that the terms democracy and politics, based on place, territory, border, and language, do not exist in a "relation of exteriority" to the technology that "one would want to be able to critique in their name." Politics is not on "secure ground" from which this technology could be analysed and its (political) consequences assessed because politics

[25] Derrida's predictions of migration on an ever-greater scale, increased flows of refugees, new forms of terrorism, and so forth have hitherto largely been proven right. These mega-trends are an undeniable feature of the world today. Large diasporas within nation states have had a considerable impact on politics and even national identity in these states, particularly Europe. And, as Derrida predicted, migration has also led to backlashes and social unrest.

itself is dependent upon the deconstructive and critical process to which teletechnology always already subjects it. The fact is that teletechnology not only "displaces places," it dissolves the very bond that connects the political and the local.

For Derrida, in other words, it is important not only to politicize technology, that is to say, to situate it in a political, social, and global context so as to determine its political, social, and global consequences. Because it is precisely *this* technology that, for better or for worse, is transforming the very concept of the political—and that charges us with the necessary imperative of thinking the political beyond politics or the democratic beyond democracy, what he calls *la démocratie à venir*.

7.4 Violation of the Inviolable: Surveillance and Control

As we have seen, the transformation of the political by modern technology has resulted in the borders of the nation state being gradually erased. Yet technology erases not only the borders of the nation state but also the boundary between the political and the non-political, the political and the private, political space and the home. "We are no longer at home!" is Derrida's dramatic diagnosis in *De l'hospitalité* (1997) with regard to the external threat to the integrity of the home at a time when both state authorities and private actors theoretically have the technological capacity to violate the integrity of every home through surveillance and control. The integrity of our homes is violated when we are deprived of the means of deciding for ourselves who we want to allow across the threshold, and when. As our homes, which ought to be inviolable, are increasingly constituted by networked technological apparatuses, state intervention therefore becomes a "violation of the inviolable."[26]

This violation of the inviolable is, I argue, a central theme of *De l'hospitalité* that has received next to no attention. The book opens with a long investigation of the classical concept of "foreigner" (Greek: *xenos*) based on close readings of Plato (*Crito*, *The Sophist*, *The Statesman*, and *Apology of Socrates*) and Sophocles (*Oedipus Rex*). Derrida takes his point of departure in the French word *étranger*, which also means "foreigner." Like *étranger*, the Greek word *xenos* denotes both a stranger and a person of foreign extraction. Derrida shows that the foreigner in ancient Athens enjoyed certain rights, such as the right to go to court to make their case. The foreigner, *xenos*, had a name, a surname, and was not the absolute other, the absolutely excluded barbarian or savage; on the contrary, *xenos* was part of a social context defined by relations between men bound by a pact of hospitality (*pacte d'hospitalité*) involving mutual rights and obligations. This foreigner, who has the right to hospitality, is part of the cosmopolitan tradition whose most significant expression is Kant's *Zum ewigen Frieden.*[27]

---

[26] Derrida, *De l'hospitalité*, p. 51.

[27] Ibid., pp. 11–31.

And yet, Derrida asks, what does "stranger" mean *today*? Concepts that we have inherited from tradition are often regarded as fundamental, natural, self-evident, and unchanging axioms. Such is not always the case, however, as the meaning of concepts is often affected by changes in the wider world: "It is often technopolitical-scientific mutation that obliges us to deconstruct; really, such mutation itself deconstructs what are claimed as these naturally obvious things or these untouchable axioms."[28]

Technology thus transforms not only politics but also our inherited concepts. How, then, ought we to understand the concept of *xenos* that confers the right to hospitality in our own digital age? Derrida's thesis is that technology, among very many other things, "completely transforms the structure of so-called public space." In Sophocles's drama, Oedipus and Antigone, as *xenoi*, have a mutual and equal relation with other *xenoi*, a relation based on the right to hospitality. But, Derrida asks rhetorically: "How would Sophocles' semantics, for example, have held up in a public space structured by the telephone, the fax, e-mail, and the Internet, by all those other prosthetic apparatuses of television and telephonic blindness"?[29]

Hospitality presupposes that we draw a line between what is ours and that which is foreign, our own home and the world outside, *the private* and *the public*. In principle, it should be possible to distribute ordinary private letters and other private consignments within a country or to another country without their being checked. In principle, it is forbidden to read or seize them. In principle, the same goes for all teletechnology, including the Internet: "Censorship, telephone tapping, interceptions, in principle represent either crimes or acts authorized only for reasons of State, of a State responsible for the integrity of the territory, for sovereignty, for security and national defense." But what does it mean today for a state to regard itself as authorized to intervene in the private sphere, in the home? "I assume," says Derrida, without knowing for sure, "that the argument by which this state intervention claims to be justified is the allegation that the space of the Internet is in fact not private but public, and above all has a public accessibility (nationally or internationally) greatly exceeding, in its usage, in its resources, that of 'porn' links by phone or video network."[30]

The dramatically increasing availability that has resulted from the introduction of digital technology makes it necessary once again to define the boundary between public and private, public space and the home, at the same time as this boundary is becoming increasingly fluid and porous. In our era, the boundary between public and private is being drawn into a legal and political turmoil, a continual process of reconstruction that challenges both existing laws and established norms. From the moment a state considers itself authorized to monitor and control exchanges of information that users regard as private but that the state monitors and controls, because exchange of information in the eyes of the state passes through public space

---

[28] Ibid., p. 45.

[29] Ibid., pp. 45–47.

[30] Ibid., p. 49. Translated updated slightly to reflect modern language usage.

and is therefore generally available, from that moment the very basis for hospitality has been removed. If the state thus uses modern technology and the Internet to intervene in whatever constitutes the private sphere, the home, "then the intervention of the State becomes a violation of the inviolable, in the place where inviolable immunity remains the condition of hospitality."[31] In other words, hospitality is the precondition for a home, just as the inviolable immunity of the home and the individual's authority over that home form the preconditions for hospitality.

State intervention by means of digital technology represents a threat not only to the integrity of the home but also the very integrity of the self. By associating the home with the self or with self-identity, Derrida creates a conception of the home as based upon the freedom of the self to decide independently whom to invite in as a guest.[32] Whenever that home is violated, he argues, we should anticipate a reaction, an increased suspicion, targeted not at the foreigner as such but at the anonymous power of technology, which threatens the very possibility of hospitality. It is only when we are masters and mistresses in our own homes that we can exercise hospitality in its traditional sense.[33] For Derrida, there is no hospitality without authority over one's own home.[34] Paradoxically, the right to hospitality for private individuals and families, that is to say, the right to filter, select, and thereby also exclude whomever one wants, if necessary by force, is regarded as something that can only be exercised and guaranteed on the basis of public or state rights. What does it mean that the state cannot "guarantee or claim to guarantee the private domain (for it is a domain), other than by controlling it and trying to penetrate it to be sure of it?"[35]

While controlling the private sphere may naturally seem negative and oppressive, the state can in turn claim to be protecting the private sphere if for no other reason than for the purpose of facilitating better communication and increased dissemination of information. An even more important reason for the state's growing need for control, however, is that the accelerating development of technology is rapidly increasing the importance and power of the private sphere, not only within its own national territory but also far beyond the borders of the nation state. Because the state is quickly becoming smaller and weaker relative to private structures of power, it is exerting itself to monitor and control the very thing that is slipping between its fingers.[36] This is primarily visible in the way that the security forces are being equipped with ever more powerful communications and information systems, enabling not only traditional wiretapping ("that is increasing every day even if, technologically, it cuts a somewhat archaic figure") but also surveillance of email traffic and the cloning of mobile phones by duplicating their contents. Additionally,

[31] Ibid., p. 51.

[32] Hospitality naturally does not require that the host welcome the guest, but it is the host's decision as to whether the guest should be invited to enter.

[33] Ibid.

[34] Ibid., p. 53.

[35] Ibid.

[36] Ibid., p. 55.

the credit card ("the ultimate identity card") is one of the security forces' most important tools.

Although the state today has considerably more sophisticated means of surveillance at its disposal than when Derrida wrote these words, it is nonetheless difficult not to agree with his contention that "current technological developments are restructuring space in such a way that what constitutes a space of controlled and circumscribed property is just what opens it to intrusion."[37]

Yet the openness of the home to the larger world is nothing new. The precondition for a home being a home in the first place is that is has "an opening, a door, and windows, a passage to the outside world." But it is only when I can invite in whomever I want, when I am master or mistress in my own house, that I can show hospitality. Conversely, such hospitality is a precondition for the home: "The monad of home has to be hospitable in order to be *ipse*, itself at home."[38] In other words, there seems to be a *double bind* between home and hospitality. While it has doubtless always been so, what is new is the massive increase in both the number of technological apparatuses and their accessibility to the world outside. In turn, this is a consequence of the fundamental similarity between the private network and the security network of state surveillance: "Their shared technology makes it impossible for the two spaces and the two types of structure to be mutually impermeable."[39]

According to Derrida, this "absolute porosity," this *limitless* possibility of accessing digital apparatuses in the home that are actually intended for the storage and communication of secret information, can be formulated as a law: "the more you encode and record in figures, the more you produce of this operational iterability which makes accessible the secret to be protected." He therefore draws the conclusion that this paradoxical law, "the perversion of this state violence or this right," is "in truth virtually inevitable." In fact, it not only eradicates the boundary between private and public, it makes the home (as a precondition for hospitality) impossible. This state "machine" renders impossible hospitality itself, the ancient right to hospitality that it ought, in fact, to be guaranteeing.[40]

How would Derrida have reacted to the revelations by American intelligence analyst Edward Snowden in the *Guardian* and *Washington Post* in June 2013 that the United States' National Security Agency, the NSA, has the capacity to monitor all of the world's internet traffic by means of a secret computing system called Prism? Surveillance by the NSA is nothing new of course—the agency was founded in 1952—but it was only after Snowden's revelations that the scale of its activities became widely known. Would Derrida, just like the EU and most European countries, have seen this surveillance as unacceptable, as a violation of the sanctity and freedom of private life? Or would have regarded it as practically unavoidably, as a natural consequence of the Internet's radical openness, and perhaps even as

[37] Ibid., p. 57.

[38] Ibid., p. 59.

[39] Ibid.

[40] Ibid., pp. 61–62.

definitive proof of the impossibility of true hospitality in our digital age? Naturally, we can only speculate about his reaction. Yet it is abundantly clear that, for him, it is a question not of either-or but both-and. Ultimately, the paradoxical question is how to preserve the sanctity of private life while also desecrating it in order to protect it, for example, from terrorists? We here encounter once again the unavoidable paradox or *aporia*, the very core of deconstruction. Derrida would probably have protested forcibly against the NSA's behaviour and questioned its almost unlimited and unmonitored surveillance methods. He would presumably have demanded greater transparency and called for the NSA and similar surveillance organizations around the world to follow existing laws and for them to be made subject to the strongest possible democratic controls, until the advent of a cosmopolitan world order beyond the nation state and its insatiable need to monitor its own citizens.

For their part, individual citizens can only conceal private letters and documents by storing them outside of the Internet, in external archives of various kinds, in which they will remain hard for the authorities to access.[41] And yet it is not only individual citizens who store important documents in hard-to-access archives; the state, science, and, indeed, all of civil society do this, too. For Derrida, the question of the archive is thus a political question, indeed, *the central issue of politics*, which recurs continually in his texts.

7.5 Archival Politics—The Central Issue of Politics

As we saw in the preceding chapter, the question of the archive has become the focus of greatly increased interest in recent years, not least as a result of Derrida's intervention. Historians, literary critics, sociologists, anthropologists as well as political scientists, philosophers, and others have begun to ask what exactly the word "archive" means. While they continue to disagree about the exact definition, they are in general agreement about the archive's central importance both for the world of research and for the existence of democratic society. For them, the archive represents the very foundation of all scientific research, the place where records are stored. It is also the place where the fundamental laws of the state are preserved. Accordingly, the archive is the place from which political power is exercised and legitimized. Or, as Derrida puts it in a crucial footnote at the start of *Mal d'Archive*: "There is no political power without control of the archive, if not of memory. Effective democratization can always be measured by this essential criterion: participation in and access to the archive, its constitution, and its interpretation."[42] The more democratic a society, the more its archives and its collective memory of the public, the citizenry, will be controlled. And conversely: the less democratic a society, the more its archives will be controlled by those with political power who wish

[41] Ibid., p. 61.

[42] Derrida, *Mal d'Archive*, p. 15, n. 1.

to limit or outright block the public's access to the archive. The question of the archive is thus a fundamentally political issue.

At the same time, Derrida underscores that "nothing is less clear today than the word 'archive.'"[43] Scholars who examine the concept of the archive usually proceed from the conventional dictionary definition as "a place where documents and other materials of public or historical interest are preserved."[44] Yet the concept is considerably more complex than its dictionary definition suggests. For scholars interested in *digital* archives, the word can mean everything that now exists in digital form, not merely in libraries, museums, and archives but, in principle, *anything*. The ongoing electronic revolution, which has quickly made every kind of archived information—sound, image, text, and multimedia—available online, has largely removed all spatial and temporal limitations, something that has changed and will continue to change our relation to the archive in ways that we cannot yet grasp. The archive, in the broadest sense of the term, increasingly seems like a metaphor for that which we do not yet understand about the nature of these digital collections. But it is already clear that the archive's central place in cyberspace can be explained by the fact that our existence in the virtual world presupposes our being interconnected. In this case, to break our connection to the digital archive is to die.[45]

It comes as no surprise to learn that the archive as a central concept has also acquired significance for political thought. The archive has quickly become a key political issue. For example, what does the archive mean for national consciousness, for democratic and totalitarian societies, and as a weapon in social, ethic, religious, and other conflicts? For Derrida, the archive is profoundly political; indeed, as he shows with particular clarity in *Force de loi*, it has been drawn into a violent political power game. For him, the question of a "politics of the archive" is therefore not only a central issue of politics; it is *the central issue of politics*: "This question can never be determined as one political question among others. It runs through the whole of the field and in truth determines politics from top to bottom as *res publica*."[46]

Derrida's *Mal d'Archive* (1995) has had a major influence on contemporary discussion of the archive's role in science and politics. The book may have its detractors,[47] but it is undeniably an extremely ambitious attempt by a philosopher to understand the desire to collect, systematize, and preserve human memories. Derrida's deconstructive analysis of the archive is innovative, far-sighted, and quite unlike anything else that has been written on this subject. In *Mal d'Archive*, he pro-

[43] Ibid., p. 141.

[44] *Webster's College Dictionary*, p. 70.

[45] Manoff, "Theories of the Archive", p. 10.

[46] Derrida, *Mal d'Archive*, p. 15, n. 1.

[47] One such detractor is the well-known British historian Carolyn Steedman. Having read *Mal d'Archive*, she argued that the archive is nothing of the sort and that Derrida's conception of the archive is excessively abstract and overly broad, "a metaphor capacious enough to encompass the whole of modern information technology, its storage, retrieval, and communication". (Steedman, "Something She Called a Fever", p. 1161).

poses a psychoanalytic interpretation of the concept of the archive. As we saw in the preceding chapter, his starting point is the fact that Freudian psychoanalysis provides us with a theory of the archive that builds on the conflict between two opposing forces. One is the death drive, and the other is the drive to preserve, the *archive drive*, which is directly connected to the pleasure principle. Derrida argues that what Freud describes as a death drive (or sometimes as an originary aggression or destruction drive) seeks to *destroy* the archive. It is what Derrida calls *anarchic*; it wields *archival violence*. It seeks to eradicate every trace of the archive, to quite simply eliminate the entire technological apparatus of documentation.[48] What happens in the archiving of a historical document is a negotiation between the death drive and the pleasure principle, between "Thanatos and Eros."[49]

The stakes in this struggle can be quite high. Although Derrida himself offers no examples, history is full of them. During the 1992 war between Abkhaz separatists and Georgia, four members of the Georgian National Guard threw firebombs into the Abkhazian national archive, thereby destroying a large part of the history of the entire region.[50] From Derrida's point of view, however, such destruction primarily represents not a destruction of human memories but a failure to take responsibility for the future in the here and now and for the archive as the very symbol of the future. A similar event took place in the wake of the United States-led invasion of Iraq in 2003. Without any real efforts being made to prevent it, the National Museum of Iraq, the National Library, the National Archive, and other collections were plundered and burned. A chorus of voices around the world denounced the incident as a cultural catastrophe of unprecedented dimensions. Some scholars and scientists even went so far as to argue that what had been lost was an artifactual legacy that could have formed the basis for the creation of a common culture among Iraq's many different ethnic and religious groups. Curiously enough, the United States Attorney General John Ashcroft offered a similar perspective when he explained that the United States would work to restore plundered museum artifacts "in part as a way of linking Iraq's ancient past to the task of building a new government."[51]

Both these examples relate to the destruction, for fundamentally political reasons, of priceless artifacts and records preserved in national libraries, national museums, and state archives with a view to depriving a people of its national and political identity. If these physical archives require a "politics of the archive," it raises the question of whether *all* kinds of archived information require some form of active politics of the archive. This is especially true of all image-based information. One of the most visible effects of technology's global spread is the enormous dominance of the *image* in an ever-growing number of areas. Images of various kinds are stored in the archives of the television, film, and cultural industries, on Internet servers, and in billions of computers, tablets, and smartphones. The

[48] *Mal d"Archive*, p. 26.

[49] Ibid., p. 27.

[50] Velody, "The Archive and the Human Sciences", p. 3.

[51] Manoff, "Theories of the Archive", p. 12.

extraordinary dominance of the image in our time undeniably derives from its capacity to affect, practically instantly, the individual's emotional experiences, something that has made images the most widespread tool for human communication, identification, and symbolization at a global level. It can reasonably be asked whether "the political" ought, in some way, to control the commercial archives that contain all of this image-based information (and, more generally, all digital information) that affects the individual's (cultural) experiences so directly and in real time. Is it time, perhaps, for such a radical politics of the archive? Or a politics of memory, to use Bernard Stiegler's phrase. The question is: how does Derrida's deconstruction view such a politics of memory?

7.6 Derrida Versus Stiegler: Messianic Ethics or Politics of Memory?

At the very end of his study *Derrida and the Political* (1996), Richard Beardsworth speculates about "two possible futures" for deconstruction. The first is what classical political philosophy would call a "left-wing" Derrideanism that asserts Derrida's analysis of an "originary technicity" and that develops the trace "in terms of the mediation between human and the technical."[52] This future forms the philosophical background of Bernard Stiegler's three-volume magnum opus *La technique et le temps* (1994–2001). The second possible future for deconstruction, which Beardsworth correspondingly terms "right-wing" Derrideanism, proceeds from "Derrida's untying of the aporia of time from both logic and technics" in claiming that time as *gift* is the very core of philosophical thinking. To choose this alternative, Beardsworth argues, is to allow oneself to be guided by a messianic promise that demands the mobilization of a "religious discourse" while passively awaiting an unforeseeable future, an unpredictable to-come. Yet Beardsworth immediately modifies this obvious dualism: "there is, of course, *no answer* and *no choice.*" In reality, there are no valid reasons for Beardsworth's "loose speculation" about a dualism in Derrida. As Ben Roberts has convincingly shown, Beardsworth's argument about two opposing "Derrideanisms" rests on a fundamental misunderstanding. For Derrida, originary technicity in the form of the trace is in fact *always already* open to the to-come, to the event, to the promise, something that Beardsworth imagines as necessarily coming *later* and as therefore involving a shift that Derrida has not expounded."[53] Yet Derrida's thought does not contain any such shift. If anything needs to "be expounded," it is the difference between Derrida's messianic ethics and Stiegler's "left-wing" Derrideanism and its demands for a politics of

[52] Beardsworth, *Derrida and the Political*, p. 156.

[53] Ibid., p. 155.

cultural memory intended to counteract what he sees as the harmful effects of global techno-capitalism.[54]

More precisely, Stiegler argues that the questioning concerning technology includes not only the danger of psychical individuation represented by the "culture industry" but also the threat posed to the possibility of alternative forms of technologically mediated collective individuation. The contemporary synchronization of all consciousness in the world via teletechnology and media is tending to reinforce the standardization and homogenization of cultural experience. This tendency towards homogenization threatens to damage or even destroy the opportunity to develop human beings' intellectual, emotional, and aesthetic capacities. It is precisely this threat to the possibility of successful psychical and collective individuation that prompts Stiegler to call for a new *cultural* politics of memory, a politics that not only neutralizes the economic forces driving addictive over-consumption in "hyper-industrialized societies" but that also serves as a counterweight to the culture industry's commercialization of teletechnologies. Such a politics of cultural memory emphasizes the importance of things like art and communication, creativity and active resistance, particularly towards the dominance of Anglo-American culture, all with a view to promoting a meaningful life, both individually and collectively, and avoiding what Stiegler calls an *industrialization of memory*.

It is here, on this particular issue, that the tension between Derrida and Stiegler begins to show. Put simply, the key question is whether modern teletechnologies deprive humans of the possibility of living a meaningful life and, if so, whether this warrants a new politics of memory or perhaps even new forms of technology? Or do they instead require a new messianic ethics—one that would avoid this kind of politics of memory and avoid direct political intervention, including in the marketplace—so as to keep the future open?

In *Échographies de la télévision*, Derrida notes that *différance* cannot be reduced to technology. Such a reduction would actually involve a radical break with the quasi-transcendental structure's character as *arche*-writing. For Derrida, however, technology represents not a break with, but rather a form of, *arche*-writing. For example, the media image is a form of *arche*-writing that it is always possible to remain critical of, precisely because it cannot be reduced to movement in *différance*. For this reason, Derrida's alternative to Stiegler's politics of memory becomes a call for a messianic ethics that is open to the other; a command to keep the future open, to refrain from a predetermined politics that risks hindering the unforeseeable arrival of the wholly other.

We have here, in a nutshell, the tension between left-wing and right-wing Derrideanism that Beardsworth saw as deconstruction's alternative futures: Derrida rejects, or at least problematizes, a materialist conception of history according to which technology is constitutive of the preconditions for humanity, something that would require a new politics of cultural memory. For his part, Stiegler embraces this

[54] Roberts, "Stiegler Reading Derrida", §23–37. For a more detailed discussion of Derrida's and Stiegler's differing views on the status of technology, see Chap. 4.

materialist conception of history and therefore endorses such a politics of memory. In other words, he denies that *différance* has a *quasi-transcendental* character; instead, he inscribes *différance* within an *empirical-material history* of technological supplements. According to Derrida, such a technological inscription in *différance* would erase the distance between technology and consciousness, thereby rendering impossible both deconstruction itself and any critical distance with regard to modern technology.

There is naturally a temptation, Derrida observes in his conversation with Stiegler, to reflexively advocate a politics of memory that would give everyone, or as many people as possible, access to a range of archives for education and research. But the problem with such a politics of memory is that it requires the intervention of a state, civil authority, or state system of power, which, because of the infinite amount of archivable material, must make a selection of what the state wishes to preserve. Priority, Derrida argues, will always be given to the national heritage and a particular interpretation of that heritage, one that invariably represents a part of the nation, such as a specific class, and not *all* citizens of that state—past, present, and future citizens. It is therefore necessary, he argues, to teach or urge each of us to think critically about the politics of memory: "we must practice a politics of memory and, simultaneously, in the same movement, a critique of the politics of memory."

However, this kind of critical thinking must not be limited to consideration of existing memories or archives. As we saw in the last chapter, Derrida emphasizes the necessity of focusing our thinking on technology's most surprising inventions, which come to us from the future. Yet such thinking confronts us with an almost impossible task: how to "politicize" these technological events? How to "democratize" them, how to criticize them, when we know that politics cannot situate itself outside of that which it seeks to criticize? Therefore, argues Derrida, we must be "be wary of a certain kind of politicization," namely that which since antiquity "has been governed, controlled, and limited" by the borders of the nation state. As we have seen, he is imagining the political in a new way, beyond these political boundaries, beyond politics.

At this point, it is worth reflecting on the rapid technological development that has taken place since Stiegler formulated his ideas about a politics of memory in the mid-1990s. On the one hand, the advent of the Internet and new social media, smartphones, and user-selected digital television has had obvious negative consequences in the form of violations of personal integrity, deregulation of the market for copyright and patents, the emergence of invisible commercial structures of power, and so forth. On the other hand, it may actually have contributed to what Stiegler was hoping for, namely greater individuation with regard to consumption and production, increased choice and personal power here and now compared with the previous, mechanical distribution of words and images—and all this without a massive cultural and political intervention.

In addition to Stiegler's materialist conception of history and his personal variety of technological determinism, Derrida calls into question his view that the market is dominant—that the economy's continual drive to expand, to find new markets, and to create new ways of consuming must lead to alienation, spiritual impoverishment, and a loss of faith in the future, something that Stiegler regards as the real danger posed by techno-capitalism. For Stiegler, it is not only the alienation of work but the alienation of individuation that represents the real danger when global techno-capitalism is combined with global technoculture. Derrida agrees with the need of some kind of state politics of memory, but he also points to the "possibilities of the market" (*la chance du marché*), its openness and unpredictability, and its historical ties with the Enlightenment and democracy (though he seems inconsistent here in his treatment of the classical liberal-democratic model whose claim to be the end-point of history Derrida calls into question in *Specters of Marx*).[55]

Derrida is in agreement with Stiegler about the need for a critical approach to modern, particularly image-based, technology. But where Stiegler calls for a politics of cultural memory involving direct state intervention in the market, Derrida affirms the Enlightenment idea of the importance of education and critical reflection for promoting active choice on the part of independent consumers and for resisting the market's demands for standardization and homogenization.[56]

7.7 Summary

In this chapter, I have sought to show how Derrida's various interventions in political questions support the overall thesis that I am advancing, namely that we can discern in them the outlines of a coherent philosophy of technology, a *phenomenology* of technology that uniquely expands the concept of technology to include the entire phenomenological field: technology is not only a material supplement to the human psyche, it is always a *relation*, a close relation to the human, life, time, religion, ethics—and politics.

Derrida was one of the first to explicitly connect politics and technology by using deconstructive analyses of these concepts. As his analyses show, the relation between politics and technology is *essential*. It is essential because technology transforms the concept of the political itself and forces us to look at the political in a new way. Now more than ever, technology operates by destabilizing the very *place* for politics in the traditional sense. The stable place for sovereign power, the territory of power, the place where power is territorially anchored—all this, according to Derrida, will disappear. Destabilization of this place involves a dislocation of space, a spatial *différance*, which challenges Western democracy based on national

[55] In *Échographies de la télévision*, Derrida makes the following claim: "The market is not an enemy of democracy … it is even a condition of democracy's development" (p. 96). And slightly later on the same page: "I don't know that anything ever escapes the market … nothing precedes what is called the market in the broad sense". See also Chap. 5 of *Specters of Marx*: "What would Enlightenment be without the market? And who will ever make progress without exchange-value?" (p. 220). Derrida's argument that exchange value (or market value) always already haunts use value is addressed on pages 228–31.

[56] Derrida & Stiegler, *Échographies de la télévision*, p. 78.

sovereignty. Trying at all costs to prop up this mode of exercising democratic power, which is quickly being eroded by and through technology, is therefore neither possible nor desirable. As technology erases the *borders* of the nation state—as ideas, sound, and images, capital and people, the arms and drugs trades, epidemics, and terrorism become increasingly borderless—the question of democracy needs to be the object of a radical reconceptualization that seeks not to abolish democracy but to develop it *on the basis of these preconditions*.

Derrida's techno-political thinking is radical not merely because it situates technology within a political context in order to determine its possible political consequences, but because it describes how *this* technology in particular is changing the concept of the political. This transformation holds especially true of political space. Technology is not only erasing the borders between nation states, it is also erasing the boundary between political space and private space. It is utterly transforming the structure of political space, something Derrida argues is almost unavoidable. It is because of this fundamental similarity between the private and public communications and information networks that the nation state has at its disposal the technological means to violate private space and the home through surveillance and control. This "violation of the inviolable" is not only a desecration of the sanctity of private life but a serious threat to the integrity of the self under the pretext of protecting us from terrorists and violent assailants. Ultimately, Derrida's analysis leads to the paradox or *aporia* that is the very core of deconstruction: the more we archive digitally, the more we give access to that which is to be protected. This aporetical "law" not only eradicates the boundary between private and public, it makes impossible the home that is a precondition of *hospitality*.

As fast-growing digital archives become increasingly surveilled and controlled by a state that feels insecure and wants at all costs to protect its citizens from catastrophes like that which took place on 11 September 2001, the question of the archive is becoming a central issue of politics. But political power not only seeks to gain control over the enormous state and private archives around the world, its own power and legitimacy are based on the possibility of control over its own *national* archive. Quite simply, there can be no political power without control over this archive. The technological archive is thus essentially connected to political power and the violent political power struggle—originating in social, cultural, ethic, religious, and other conflicts—that simultaneously establishes and conserves the laws that legitimate power. This is the reason why Derrida considers the question of the politics of the archive as the central issue of politics, the issue that determines the political from beginning to end.

An interesting instance of archive politics is offered by Bernard Stiegler's politics of cultural memory, which seeks to counteract the threat posed by global techno-capitalism to psychical and collective individuation. Derrida problematizes this politics of memory and its demands for comprehensive state invention. As an alternative, he advocates a messianic ethics that is open to the future, to the unforeseeable arrival of the wholly other. Openness of this kind presupposes that *différance* is not reducible to technology, something that would render impossible both deconstruction itself and any critical distance to technology. Derrida also argues against the state's absolute dominance of the market and its opportunities, openness, unforeseeability, and historical ties to the Enlightenment and democracy. While he supports a politics of memory, he also calls for a *critique* of it on the grounds that the state can choose to limit access to archived materials that it has an interest in citizens being able to get hold of. In other words, it involves opening up the kind of democracy that is governed, controlled, and limited by the borders of the nation state. It is a matter of looking to the democratic beyond democracy or to the political beyond politics.

This *beyond* is what liberates politics from identity with itself and connects it with the wholly other, the "nonnegotiable" that Derrida calls ethics. The concepts of technics, ethics, and politics are thus indissolubly interconnected in a global techno-ethico-political complex in which accelerating technological development is itself the engine of a movement towards a new world order.

References

Beardsworth, R. (1996). *Derrida and the political*. Routledge.
Derrida, J. (2003). *Voyous: Deux essais sur la raison*. Galilée.
Derrida, J. (2002a). Politics and friendship. In E. Rottenberg (Ed.), *Negotiations: Inventions and interviews, 1971–2001*. Stanford University Press.
Derrida, J. (2002b). Ethics and politics today. In E. Rottenberg (Ed.), *Negotiations: Inventions and interviews, 1971–2001*. Stanford University Press.
Derrida, J. (1997a). *De l'hospitalité*. Calmann-Lévy.
Derrida, J. (1997b). *Adieu à Emmanuel Levinas*. Galilée.
Derrida, J. (1995a). *Mal d'archive: Une impression freudienne*. Galilée.
Derrida, J. (1995b). *Archive fever: a Freudian impression*. (trans. E. Prenowitz). The University of Chicago Press.
Derrida, J. (1994a). *Politiques de l'amitié*. Galilée.
Derrida, J. (1994b). *Specters of marx: The state of the debt, the work of mourning, and the new international*. (trans. P. Kamuf). Routledge.
Derrida, J. (1992). Force of law: The "mystical foundation of authority". In D. Cornell, M. Rosenfeld, & D. Carlson (Eds.), *Deconstruction and the possibility of justice*. (trans. M, Quintenance). Routledge.
Derrida, J. (1988). In G. Graff (Ed.), *Limited Inc.*. (trans. S. Weber, J. Mehlman). Northwestern University Press.
Derrida, J. (1967). Violence et métaphysique. In *L'écriture et la difference* (pp. 117–228). Seuil.
Derrida, J., & Ferraris, M. (2001). *A taste for the secret*. (trans. G. Donis). Polity Press.
Derrida, J., & Stiegler, B. (2002). *Echographies of television: Filmed interviews*. (trans. J. Bajorek). Polity Press.
Derrida, J., & Stiegler, B. (1996). *Échographies de la télévision*. Galilée.
Manoff, M. (2004). Theories of the archive from across the disciplines. *Libraries and the Academy, 4*(1), 9–25.
Roberts, B. (2007). Stiegler reading Derrida: The prosthesis of deconstruction in technics. *Postmodern Culture, 16*(7).
Steedman, C. (2001). Something she called a fever: Michelet, Derrida, and dust. *American Historical Review, 106*, 4.
Volody, I. (1998). The archive and the human sciences. Notes towards a theory of the archive. *History of the Human Sciences, 11*, 4.
Webster's College Dictionary (1999). : Random House.

Chapter 8
Religion and Technology

Religion today allies itself with tele-technoscience, to which it reacts with all its forces—
Jacques Derrida, *Faith and Knowledge*

8.1 Introduction

What is the relationship between religion and science? Today, many scientists, philosophers, and non-believers, particularly in the industrialized world, regard these two domains as almost incompatible. This incompatibility applies equally to the relation between religion and technology since technology is built on science and science, especially nowadays, cannot be produced without technology. Conversely, representatives of religion have often manifested a genuine antipathy towards technology, which is partly explicable as a natural reaction to the destructive forces of technology but can also have its basis in religious fundamentalism, religious archaism, or notions of spiritual purity. Whatever the case, there is a prevailing view that religion and technology are entirely separate domains with no points in common.

In his many writings on religion, Derrida seeks to show that religion and technology are not independent of each other but always already intertwined in a close relation. Before going into the character of this relation, we need first to recapitulate some central ideas in his philosophy of technology.

In the preceding chapters, I have sought to show how Derrida's thinking is permeated by reflections on technology. Although he never wrote an entire book on the philosophy of technology (something that is also true of other philosophical subjects that were key for him), I argue that his thinking about technology is both acute and innovative. I have tried to sketch the outlines of a coherent philosophy of technology in Derrida's thinking that is not merely characterized by a fundamental philosophical approach or attitude but that also engages with technology itself, not primarily as a philosophical idea but in terms of different technologies' relation to human beings and the world. Derrida's philosophy of technology, whose central ideas I will now try to summarize in a few pages, is thus both a *philosophy* of technology and a philosophy of *technology*.

B. Sjöstrand, *Derrida and Technology: Life, Politics, and Religion*,
https://doi.org/10.1007/978-3-030-83407-4_8

Derrida's *philosophy* of technology is firmly anchored in a philosophical tradition that seeks to understand modern technology on the basis of a hermeneutic-phenomenological perspective. This does not mean that he always agrees with the views of its representatives; on the contrary, his deconstructive analyses of other philosophers' ideas about technology and science are sceptical and problematizing. And yet this lifelong and productive dialogue with Husserl's and Heidegger's texts forms the basis of Derrida's own reflections on technology.

The phenomenological leitmotif of Edmund Husserl's texts was to go "to the things themselves!" Derrida might be said to have taken this injunction literally. What holds for Husserl's phenomenology holds equally true for Derrida's philosophy of technology: go to the things themselves! This does not mean that he views phenomenology as a method with whose help the "things themselves" can be described in a more originary way than is currently possible using scientific methods; rather, he views phenomenology the way Husserl himself understood it, at least in his early works, namely as a tool for analysing *the relation* between human beings and their world. The "things themselves" thus seem to refer not to things in the world but rather *the relation between human beings and the world.*

In this phenomenology, which arose as an attempt to overcome the tensions between idealism and realism in nineteenth-century theories of knowledge, Husserl's concept of the *intentionality* of consciousness played an important role. Like his followers, the later Husserl came to realize that phenomenology, as a philosophy of transcendental consciousness, was only one aspect of the relation between human beings and the world, and perhaps not even the most relevant one at that. Moreover, the world cannot be understood as a collection of objects of scientific knowledge but must, rather, be seen as something in which people live: a *lifeworld.* Husserl's philosophy of consciousness was thereby broadened into an analysis of the relation between human beings and their world in a general sense.

When Derrida claims that this relation is fundamentally technological in nature, he is precisely starting from Husserl's conception, as set forth in "The Origins of Geometry" and elsewhere, of a naïve, pre-theoretical lifeworld, an originary practical or technical activity in the world. What both Husserl and Derrida show is that some kind of technical activity always already precedes objective knowledge of the world. For both thinkers, writing—graphic inscription—is the technology that guarantees this scientific objectivity. In written form, an originary meaning can be objectively preserved and handed down from generation to generation, to be incorporated into a shared bank of experience that future scientists can draw on and expand. In this way, writing is the opening of a tele-communication, a technological tool that *mediates* objective knowledge across a distance in space and time.

But Derrida here departs significantly from Husserl's transcendental phenomenology. Writing is not merely *one* empirical or technological tool at the subject's disposal but the very precondition for that subject to be able to make claims to objective truth in the first place. Yet this precondition can hardly be empirical in nature since it is logically impossible to acquire knowledge of something that is itself a condition of possibility for that knowledge. Writing must therefore be understood in two ways: as writing in the usual, empirical sense, i.e. as technology; and as a subjectless condition of possibility for knowledge and meaning, i.e. the quasi-transcendental that Derrida subsequently termed "arche-writing." For this reason, I have chosen to describe his philosophy of technology as a *phenomenology of technology* in which writing in the general sense, rather than the transcendental subject, is constitutive of consciousness.

Yet Derrida's philosophy of technology is not only removed from this philosophy of transcendental consciousness, it is also incompatible with the opposite position, namely, a purely empirical approach in which the primary task of philosophy of technology is the empirical study of concrete technologies and their interaction with society more broadly. While this empirical approach accords concrete technologies the attention they undeniably warrant, it does not in itself constitute an alternative to classical philosophy of technology. The reason is that empirical studies of technology barely address the phenomenological, hermeneutic, or existential questions posed by classical philosophy of technology.

By contrast, Derrida's philosophy of technology has its basis in the phenomenological and hermeneutic ideas of Husserl and Heidegger even as it broadens the concept of technology to include the entire phenomenological field: there is no possibility of shielding thought from concrete, empirical technology. For Derrida, thought and technology are not opposites: thought is always already *contaminated* by technology—by the technological apparatuses and artifacts that confront us at virtually every turn in our daily lives.

Derrida's philosophy of *technology* therefore sees its task as the interpretation of human beings' relation to the technological apparatus. It views this technological apparatus as a *supplement* to human beings. It is an external "memory" and, as such, is a necessary supplement to humans' living memory if the latter is to function at all. Technology is therefore not the technological apparatus as such but the *relation* between the human "psychical apparatus" and the technological apparatus, between the living and the non-living, between life and death. Technology is the writing that opens up in the difference between internal and external memory, between what Plato called *mneme* and *hypomnesis*. Derrida thus defines technology in terms of relations between psychical process and different technologies for the inscription and storage of all text, from clay tablets and papyrus rolls to books and digital media. He summarizes the telecommunicative technologies of our era by referring to a gigantic "teletechnological apparatus" that includes not only media in the form of the press, radio, television, and film but also digital archival systems, systems for mobile communication, and a global infrastructure in the form of industries, experts, treaties, and labour practices.

This teletechnological apparatus is today characterized by the digitization of practically everything that can be digitalized, something that is dramatically reducing time and space on every level and ratcheting up the pace of contemporary social life. It is, moreover, virtually *borderless*: it does not recognize national borders and it mediates information and knowledge across the world in effectively unrestricted ways. While the current concept of democracy "is determined, controlled, and constrained" by the borders of the nation state, Derrida predicts that the teletechnological apparatus will blur these boundaries and perhaps, in some distant future, even dissolve them entirely. Demands for national sovereignty and claims for citizenship and democratic rights, all of which have their historical legitimacy inscribed within the framework of territorial borders, will perhaps lose their meaning as a result of the impact of modern technology: "What the accelerated development of teletechnologies, of cyberspace, of 'the virtual' is producing is a practical *deconstruction* of the traditional and dominant concepts of the state and citizen (and thus of the political) as they are linked to the actuality of a territory."[1]

It therefore follows that there is a close relationship between technology and the political. According to Derrida, this relationship is one of autoimmunity. The concept of autoimmunity—which he used with increasing frequency during the last two decades of his life—seems to have been the final variant of what for more than forty years he called *deconstruction*. This terminological change is important for understanding not only his later works but also his entire oeuvre. By foregrounding a term taken from the life sciences that describes the process by which an organism almost suicidally turns upon its own defence system, Derrida was able to show how all political identities, from the local community to the nation state, are subordinate to a process that both threatens to destroy them and makes it possible for them to continue living. This autoimmunity is undoubtedly an expected and far more disturbing way of understanding deconstruction, even if the difference between them is considerably less than one might initially assume. Since Derrida, as we saw in the discussion at the start of Chap. 6, defines deconstruction as an unconditional welcoming of alterity, there is always a risk that this welcome will fundamentally jeopardize one's own identity. We might say that deconstruction is always open to a perversion of autoimmunity. Welcoming an event can therefore just as easily lead to something evil as to something good; it can represent a threat as much as an opportunity, death as much as life.

While deconstruction has often been erroneously regarded as a "method" or "strategy" for analysing text or concepts, "autoimmunity" seems to be the name for a process that necessarily operates virtually everywhere, particularly at the heart of all political identity. "Autoimmunity" is not primarily a method or strategy of theoretical textual analysis but a kind of "death drive," to use Freud's term, a death drive that is an integral part not only of the human psychical system but of all nation states and political institutions.

This autoimmune process is driven by technological development. When Derrida, in what seems to be his fundamental political thesis, claims that technological development produces a *practical* deconstruction of the political, it means that the relation between technology and politics is not only essential but autoimmune. The erosion of social, cultural, and national boundaries by accelerating teletechnology inevitably brings with it a destabilization of, and perhaps even a deadly threat to, every kind of sovereign identity. The ever-faster transformation of technology itself appears as a powerful driving force on the road to a world defined by democracy and justice even as it can give rise to counter-reactions in the form of religious

[1] Derrida & Stiegler, *Échographies de la télévision*, p. 45.

fundamentalism, national archaism, ethnic separatism, and so forth, as we saw in Chap. 6. These counter-reactions are often associated with destructive uses of the very technologies to which they are opposed. Uncanny examples include the use of commercial passenger jets as deadly weapons, the proliferation of chemical, biological, and nuclear weapons of mass destruction and their exploitation by terrorists, or the threat of cyberwarfare for the purposes of instantly disabling global communication systems and archives.

These examples illustrate the way in which technology produces an autoimmune process that, by erasing social, cultural, and national borders, promotes positive values such as democracy and justice while simultaneously constituting a deadly threat to all political sovereignty. For better or for worse, technology is transforming the very concept of the political and charging us with a necessary imperative: to think the political beyond politics or the democratic beyond democracy, what Derrida calls *democratie à venir*.

Even so, when all is said and done, when we have described Derrida's philosophy of technology as a phenomenology of technology and technology's relation to ethics and politics as an autoimmune relation, there is nonetheless something essential lacking in our understanding of his philosophy of technology. What is needed is a deeper analysis that does justice to his ideas about the relation of technology to *religion*. As we will see, religion is an aspect of the question concerning technology that provides the key to a more comprehensive understanding of his view of technology.

8.2 Derrida on Religion

Many commentators claim to discern a religious or theological discourse in Derrida's writing. For example, Jürgen Habermas argues that deconstruction contains a faith and a mysticism that are fundamentally irrational. In *The Philosophical Discourse of Modernity* (1987), he dismisses deconstruction as a religious, mystical, and even Talmud-inspired mode of thought, arguing that Derrida's texts can hardly be considered part of philosophy as a rational discipline.[2] Other scholars, such as Hent de Vries och John D. Caputo, refer to a religious turn in his later writings, a growing interest in religion that reveals a kind of "atheism" within the religious tradition.[3] Reference is here usually made to *Donner la mort*, in which Derrida refers to deconstruction as "the *possibility* of religion without religion," a religion that does not need the religious "*revelation of an event.*"[4] But others argue, conversely, that no such religious turn exists in his writing. Martin Hägglund even

[2] Habermas, *The Philosophical discourse of Modernity. Twelve Lectures*, p. 182.
[3] Caputo, "The Return of Anti-Religion", p. 109.
[4] Derrida, *Donner la mort*, pp. 74–5.

claims that a "radical atheism informs his writing from beginning to end."[5] In fact, Derrida's deconstructive analyses of religion and theology are acute and penetrating: they investigate and question the traditional divide between faith and knowledge, a divide that is the very precondition of all mysticism and all theological thinking.

In order to understand what Derrida means by the paradoxical phrase "religion without religion," we need first to say a few words about his own religious background. He was born a Sephardic Jew but grew up in a largely Muslim society in Algeria, which was a French colony at that time. He lived almost his entire adult life in the Christian, and specifically Catholic, country of France, for which reason his intellectual milieu was European and Christian. But in addition to these Jewish, Muslim, and Christian heritages, which were to shape him in various ways, he also cultivated a degree of "atheism" in the communities of which he both was and was not a part. For example, he declares in *La contre-allée* (1999) that "my atheism develops in the churches, all the churches."[6] Although Derrida did not develop a theological system, we might say that the position he adopts seems compatible with an intraworldly Christian theology, a kind of Christian atheism that has a number of similarities with negative theology.

But Derrida nonetheless has reason to doubt that such an intraworldly Christian theology can be formulated. While it is true that he often refers to *la différance* in terms of negative theology ("neither this nor that"), it does not mean that this "non-concept" beyond being is divine in nature. Negative theologies regard God as something transcendent, while *la différance* is a quasi-transcendental condition of possibility of knowledge. Moreover, he suspects that even the most negative of negative theologies equates God with total presence, thereby entangling itself in what he regards as the fundamental prejudice of Western philosophy—the metaphysics of presence.[7] Negative theology is thus hardly a viable path towards an intraworldly theology.

While Derrida regards negative theology as a dead end on the path towards an intraworldly theology, he is far more optimistic about the possibilities offered by a deconstructive theology. The aim of the latter would be to "liberate theology from what has been grafted on to it, to free it from its metaphysico-philosophical super ego, so as to uncover an authenticity of the 'gospel.'"[8] Yet we should not regard this deconstruction as some externally derived investigation of theology; rather, deconstruction has from the very start been working within theology itself by questioning Aristotelianism or Thomism, that is to say, an entire theological institution that has clearly concealed or masked an authentic Christian message.[9]

---

[5] Hägglund, *Radical Atheism*, p. 1.

[6] Derrida, *La contre-allée*, p. 99.

[7] Hart, "Religion", p. 56.

[8] Derrida, "Deconstruction in America", p. 12.

[9] Ibid., p. 12.

When Derrida here refers to the authentic Christian message of the gospel, he evidently has Paul the Apostle and the Pauline epistles in mind. As Hans Ruin has recently shown, Derrida's texts can be read as an implicit mode of thinking in the Pauline tradition, even as they should be understood as an important critique of the way in which that tradition is administered today.[10] While Derrida may not have written a single text that exclusively addresses the Pauline epistles, this does not change the fact that his entire oeuvre is permeated by themes that were also central to Paul: bodily and spiritual circumcision, Jewish and Greek spirituality, the spiritual as such, hope and justice, universalism and cosmopolitanism, and even the very idea of deconstruction as an intellectual strategy. Deconstruction is thus not only a strategy that can now be applied to Pauline texts, it can also be seen, from a certain viewpoint, as an integrated part of the Pauline intellectual legacy. As such, it can and should be an active force within theology itself.

While deconstruction *today* should be an active force within theology itself, in reality it has been an active force in Christianity throughout its entire history. There is now a consensus that the very word "deconstruction" is partly a derivation from Luther's *destructio*, which in turn is a reply to Paul's words in 1 Corinthians 1:19: "I will destroy the wisdom of the wise; the intelligence of the intelligent I will frustrate," a direct quotation from Isaiah 29:14. Luther's *destructio* of the Catholic Church's pomp and splendour and replacement by a more originary faith anticipates Heidegger's *Destruktion* of the history of ontology and return to a more originary, pre-Socratic experience of Being. As early as the 1920s, Heidegger gave lectures on Christianity, Kierkegaard, Luther, and, above all, Paul, whose epistles he gave a secular-philosophical, demythologizing interpretation, which was subsequently further developed by the German theologian Rudolf Bultmann.[11] Inspired by the Christian tradition, Heidegger's *Destruktion*, along with *Abbau*, is now considered an immediate precursor of the term deconstruction.

A certain kind of deconstruction is always already theological, and its theology emerges from the Reformation. At the same time, it is easy to see how a deconstructive theology of a different kind might call into question the Protestant doctrine of a return to a pure origin and an unmediated relation to God. There is no *one* deconstruction, says Derrida, and—we might add—nor is there *one* deconstructive theology.

Derrida's interest in negative theologies was mainly concentrated to the mid-1980s and early 1990s, after which he became more concerned with the question of religion today and how we should re-evaluate concepts such as faith and salvation, good and evil, life and sacrifice. His late thinking can be summarized in the paradoxical phrase "religion without religion." The formula "x without x" goes back to Augustine, who used it to convey God's absolute transcendence. Derrida adopted the formula from Maurice Blanchot—who had, in turn, borrowed it from Levinas's famous "relation without relation" in *Totalité et Infini*—precisely in order to reject the transcendence that Augustine sought to assert.

[10] Ruin, "Circumcising the Word: Derrida as a Reader of Paul", p. 73–5.

[11] Ibid., p. 78.

Derrida's definition of deconstruction in *Donner la mort* as a mode of thought that "repeats" the possibility of a religion without religion means that his thinking returns repeatedly to religion, to Jewish, Muslim, and Christian themes, problems, and traditions: faith and knowledge, the sacred and the impossible, life and sacrifice, and, perhaps most importantly of all, the messianic. As is well known, he mentions at several points in later works such as *Specters of Marx* that deconstruction can even be defined as a "messianism without messianism" in that it is radically open to that which comes from an unforeseeable future.

The so-called "religious turn" in Derrida's late writings is undoubtedly one of the most controversial issues in contemporary Derrida scholarship, and his commentators have advanced diametrically opposed interpretations of its significance. On the one hand, scholars such as Kevin Hart, Hent de Vries, and, above all, John Caputo have seen this religious turn as the logical consequence of something that forms the very core of his thinking, namely the messianic requirement of radical openness to the wholly other. On the other hand, Richard Beardsworth, Bernard Stiegler, and, most recently, Martin Hägglund have attacked this religious turn as a spurious attempt to inscribe a transcendent or theological discourse within this core. Hägglund even describes deconstruction as "radical atheism," which means that its origin in *différance* and its critique of all forms of metaphysics of presence necessarily lead to a radical, non-religious, and non-theological position. Where the former grouping in their respective ways support the view of deconstruction as a "religion without religion," the latter attack any such reading, which they see as a denial of the necessary empiricism or materiality *within* alterity.

This heated debate, whose main protagonists are John Caputo and Martin Hägglund, has many similarities to the debate we examined in the previous chapter, namely Richard Beardsworth's speculations about the two, ostensibly incompatible futures for Derridean deconstruction. Either embrace a "left-wing" deconstruction that takes its staring point in technology and views technology as constitutive of the human—a future exemplified by the philosophical background to Bernard Stiegler's *La technique et le temps*—or profess allegiance to a "right-wing" deconstruction, which champions the late Derrida's quasi-transcendental or even theological discourse of a messianic promise, passively awaiting an unforeseeable future. As we showed in the last chapter, Beardsworth's dualism rests on a misunderstanding, since technology is *always already* open to the future, to the messianic, to the promise. Towards the end of this chapter, I seek to show that Beardsworth's, Caputo's, and Hägglund's interpretations of Derrida all partly rest on a misunderstanding of the deeper impulse at work in Derrida's writing. It is that we cannot choose between a "religious" and a "technological" deconstruction because religion and technology are always already essentially connected.

The question of religion's relation to technology is far from new. It has been on the philosophical agenda since at least the beginnings of modernity, even if it has traditionally been described in terms of an opposition between religion and science, faith and reason, faith and knowledge. In our era, this opposition can be discerned in debates about creationism and Darwinism, stem-cell research and genetic engineering, and, above all, in the battles over the publication of what many people regard as offensive representations of religious symbols. While Derrida does not specifically address these problems in his many writings on religion, it should be possible to approach virtually every problem in philosophy of technology—the relation between religion and technology, for example, or the relation between religion and technoscience and the globalization process that it is driving—in a new way in the light of the provocative analyses and arguments that Derrida advances in these works.

This is not the place for a more systematic consideration of Derrida's many writings on religion, which have already been analysed far more thoroughly than is possible in the limited space available here.[12] However, a handful of important works will serve to give an indication of his prolific output on this topic: *Glas* (1974), "Circumfession" (1991), *Voiles* (1998), *Donner la mort* (1999), and, above all, *Acts of Religion*, which was originally published in English in 2002 and comprises a collection of key essays on religion, including what is undoubtedly Derrida's most important text on this subject, which we will now consider more closely.

8.3 Religion as Abstraction

In February 1994, Derrida took part in a conference on Capri devoted to the question of the nature of religion and its role in the contemporary world. Derrida's essay *Foi et savoir: Les deux sources de la "religion" aux limites de la simple raison* (1996) is a revised and extended version of the reflections that he offered on that occasion. "Faith and Knowledge," to give it its English title, is a dense, not easily parsed, and unusually synthetic text in which Derrida shows his profound engagement with the question of the nature of religion in general and of religion's two "sources" (to quote his essay's subtitle) while also calling into question the forms taken by religion in the present. It addresses topics that one would expect to find in a study of religion, such as the nature of revelation, faith, salvation, prayer, sacrifice, God, sanctity, and so forth, but also less predictable subjects such as teletechnology, technoscience, telecommunication, media, politics, ethics, sexual violence, etc.

"Faith and Knowledge" can be read as an introduction to the question of religion is Derrida's oeuvre as a whole. It summarizes large parts of his previous writings and anticipates much of his subsequent work. It is, as Michael Naas rightly observes in his erudite, insightful, and highly readable study *Miracle and Machine* (2012), "an absolutely crucial essay … for understanding not just Derrida's work on religion but his work as a whole."[13]

The central questions in this important work concern the relation between faith and knowledge, religion and science, and religion and technology ("Faith and

[12] See, for example, Michael Naas, *Miracle and Machine*, p. 338 n.6, which includes a list of works on Derrida's ideas about philosophy of religion.

[13] Naas, *Miracle and Machine*, pp. 1–2.

Knowledge" is without doubt one of Derrida's most thorough examinations of contemporary technology and science). In it, he ascribes a fundamental importance to the *autoimmune* relation between religion, technology, and science, particularly the paradoxical fact that many so-called extremist or fundamentalist religious groups reject technological-scientific modernity precisely by creatively appropriating and using its *technoscience*.

First launched within the field of science and technology studies, technoscience is a modern concept which emphasizes that there are no pre-existing boundaries between what constitutes technology and what constitutes science. As Michael Naas rightly points out, this emphasis upon the *performative*, performance, is what leads Derrida in "Faith and Knowledge" to refer less to science than to technology or (tele-)technoscience.[14] Derrida makes explicit reference to "science as technoscience, as critical history of the production of knowledge, of knowledge as production, knowhow and intervention at a distance, teletechnoscience that is always high-performance and performative by essence, etc." (§29). In his essay, Derrida seeks *both* to explain the autoimmune logic behind religion's simultaneous rejection and embracing of technoscience *and* the global consequences of this logic. These explanations rest in part on the claim that an originary or elemental (but non-dogmatic) faith is a precondition for religion as well as for technology and science, a statement that may shed new light on many contemporary questions about religion's role in society, the desirability or possibility of a truly secular state, or the very possibility of differentiating clearly between religion and technology or religion and science.

Before examining Derrida's reflections on the question of religion and its relation to technology and science more closely, we must first understand the background to "Faith and Knowledge." In some of its early chapters, Derrida considers the very circumstances and conditions under which his essay came to be written, something that is important in this context because these circumstances and conditions substantially determined the contents of the essay itself. As Naas notes, these circumstances are, in fact, completely crucial to an understanding of its themes and conclusions.[15] I will therefore briefly summarize the timing and location of the conference, its format and language, and the political, philosophical, and religious background of the delegates as they are presented in the essay.

Derrida likens the framework of the conference, which gathered delegates on the relatively sparsely populated island of Capri in late February 1994 to discuss "religion," isolated from the rest of the world in a kind of retreat, a place for spiritual exercise, a kind of monastery for a small religious community or sect, which—to his clear disappointment—consisted entirely of men.[16] No women were invited—"*Not a single woman!*," remarks Derrida pointedly—to discuss a subject of such

---

[14] Ibid., p. 90.

[15] Ibid., p. 23.

[16] The other conference delegates were Hans-Georg Gadamer, Gianni Vattimo, Eugenio Trias, Aldo Gargani, and Vincenzo Vitiello.

tremendous importance to women as religion. He was equally disappointed with the delegates' religious backgrounds, noting that "*we represent and speak four different languages, but our common 'culture,' let's be frank, is more manifestly Christian, barely even Judaeo-Christian. No Muslim is among us, alas.* (§5)."[17] The delegates at this gathering, this retreat, were thus all men, European men—"we **Europeans**" is the phrase Derrida uses—European men with a visibly Christian background. For him, women and Muslims are important symbols and themes that were excluded from the Capri conference but that he would return to at the end of his essay.

The four languages spoken at the Capri conference were all European: Italian, French, Spanish, and German. "Faith and Knowledge" discusses at several points the question of *language*, particularly the relation between religion, the nation state, and language or national idiom. As Derrida affirms in an early paragraph (§4), language or writing is the very heart of all *revelation* and all *religious conviction* and can never be separated from a specific social, political, or ethnic location—a nation, a people, or a country. When reflecting on or discussing religion, we must always connect it with the national idiom, with blood, earth, place, citizenship, the nation state, etc.—a theme that Derrida examines in several books, including *Psyché* and *Specters of Marx*, and that was discussed in detail in the preceding chapter. It is in this context that he introduces the unwieldy neologism *mondialatinization* or "globalatinization" in order to underscore the fact that Latin is the basis for the concept of religion, its definition, and its global dissemination, a thesis that leads him to the thought that, strictly speaking, perhaps only Christianity can be characterized as a "religion" (§2 and §5).

While all these preconditions for a discussion of religion—time, place, format, religious affiliation, and national idiom—from the very beginning exclude certain perspectives, such as those of women, Muslims, and non-Europeans, the delegates seem to be in agreement on at least one thing. Because the discussion is taking place within the framework of a Judaeo-Christian tradition, they share what Derrida calls a certain *pre-understanding* of the subject. Even if we cannot *know* for certain what religion is, what the word means, he observes, we nonetheless have a kind of *faith*: we think we have a pre-understanding of the meaning of the word. We believe in a minimal stability (*fiabilité*) in this word (§3). We may not know the exact meaning of the word but we at least share a kind of faith in what might be called the *horizon* of the question of religion.

This reference to a shared horizon and pre-understanding leads directly onto another precondition, another reference point. Although the delegates at the Capri conference speak four different languages, come from four different countries around the Mediterranean, and have partly differing religious backgrounds, all have

[17] "Faith and Knowledge" is written in the form of a series of aphorisms, divided into 52 sections or "paragraphs", of which the first 25 are written in *italics* and the other 26 in regular roman script. (The former seem to summarize Derrida's oral presentation in Capri, while the latter seem to correspond to reflections that he compiled after the conference.) Since the various editions of "Faith and Knowledge" contain significant differences, I have chosen to refer to the text by paragraph numbers.

been shaped by European philosophy and, notwithstanding their many differences, have been "*oriented by a certain phenomenology*" (§10) as if by a kind of magnetization. They have all been inspired by Husserl and a hermeneutics that has meant so much for the interpretation of religious texts. As a result, all of the delegates are very familiar with hermeneutical-philosophical concepts such as horizon and pre-understanding.

Politically, the delegates all share the view that religion must be separated from politics, and all have "*an unreserved taste, if not an unconditional preference, for what, in politics, is called republican democracy as a universalizable model*" (§11). While there are those opposed to the involvement of religion in political life, none of them can be thought of as sworn enemies of religion or advocates of a doctrinaire secularism. Derrida further notes that, regardless of their own religious affiliation, none of them are "*priests bound by a ministry, nor theologians*" nor "*qualified, competent representatives of religion*" or "*enemies of religion as such, in the sense that certain so-called Enlightenment philosophers are thought to have been*" (§11). Although they want to keep the political sphere free of religious involvement, none of them would definitively distance themselves from religion as such, and all of them, including Derrida, are staunch defenders of religious freedom.

So far I have summarized Derrida's opening reflections in "Faith and Knowledge" on the timing and format of the Capri conference, the languages spoken, and the delegates' gender, nationality, and political, philosophical, and religious background. While these reflections on the conference *setting* may seem somewhat peripheral, they were in fact strategically important for him. He wishes to highlight this background or *context* in order to emphasize the necessity of attending to every factor that can help to make the conference "*an effective and unique situation*" (§4). According to Derrida, every discourse must be *situated* in order to be understood; its context must be clarified in order to be understood as a unique event. In order to discuss something as abstract as religion, we must first determine the immediate context of the present, the unique situation, the unique event.

But the current context and the pragmatic preconditions of an event are not something to which Derrida merely refers or takes into consideration in his treatment of a subject; these factors are always essential for an understanding of his line of argument. This is especially true of "Faith and Knowledge," in which, on the basis of these individual contextual factors, he transforms each and every one of them into a theme of his text. Its location, the empty island as a place of abandonment, becomes a reflection upon the spatiality of revelation, upon *khora*, and so forth. Its timing becomes a series of thoughts on history and the historicity of religion, a train of ideas about the situation of religion today. The languages spoken at the conference become a line of thinking about the Latin origins of contemporary religious discourse. The absence of women becomes a discreet but unmistakeable articulation of the female voice throughout the entire essay. And the absence of

Muslims becomes a continual reminder of the role Islam must play in any serious consideration of religion in the present.[18]

The interesting thing about these themes is that they collectively support what I am arguing is the main thesis of Derrida's essay, namely the unavoidable and autoimmune relation between religion and technology. Even the format of the text is a kind of indicator of this relation: as we have seen, "Faith and Knowledge" is a series of fifty-two numbered aphorisms or "paragraphs," the first twenty-six of which are italicized and the remaining twenty-six are in roman. In the second of these paragraphs, Derrida claims that the numbering has a particular relationship to the aforementioned relation between religion and technology or the machine. He feels compelled to invent a "*discursive machine*" whose task is to say something substantial about so large and important a subject as religion within the framework of "a certain speed, at a rhythm given within the narrow limits available" (§2.33).[19]

Derrida's use of the word "machine" here provides an indication of the way in which religion today is being reshaped by technoscience, the telecommunications networks and media that now span the world. We will return to this problematic shortly. But first we need to clarify what we mean by the concept of religion. What *is* religion *today*? How can we speak of religion today? Derrida begins "Faith and Knowledge" as follows:

> *How "to talk religion"? Of religion? Singularly of religion, today? How dare we speak of it in the singular without fear and trembling, this very day? And so briefly and so quickly? Who would be so imprudent as to claim that the issue here is both identifiable and new? Who would be so presumptuous as to rely on a few aphorisms? To give oneself the necessary courage, arrogance or serenity, therefore, perhaps one must pretend for an instant to abstract, to abstract from everything or almost everything, in a certain way. Perhaps one must take one's chance in resorting to the most concrete and most accessible, but also the most barren and desert-like, of all abstractions* (§1).

Derrida begins, then, by discussing the difficulty of talking about religion today or (the real problem) "talking religion"—as if religion itself were a kind of form or language. The question is thus not how we can talk about religion without generalizing or perhaps even profaning, but how we can talk religion when the language we use, perhaps from the very beginning, is the language of religion—in other words, the Latin language that defines and has defined so extraordinarily many of our religious terms and concepts, including the concept of *religion* itself. To ask what religion *is* by using the word *religion* is perhaps already a kind of answer in itself.

Derrida further asks how we can talk about religion in the *singular* today, as if it were a pure monolith, as if the three monotheistic religions of Christianity, Judaism,

---

[18] Naas, *Miracle and Machine*, p. 39.

[19] As well as being divided into two parts and two typographic fonts, the latter part, which is titled "Post-scriptum", is further divided into one group of eleven paragraphs and another of fifteen paragraphs. As if this were not enough, Derrida creates yet another typographic level in the form of a playful alternation between normal and bold font, which collectively form a kind of complicated, not easily interpreted, but above all meaningful pattern of words and phrases, a kind of textual *invention*.

and Islam were a monolithic entity with no significant differences—let alone the issue of the world's other religions. He has a genuine respect for this issue, to the point of articulating his fear and trembling at it—originally a phrase of Paul's and now usually associated with Kierkegaard. Derrida here uses the concept of *abstraction*, which must be understood in at least two senses. First, he is asking whether we have to abstract from everything else, that is to say, to bracket everything else, and perhaps primarily reason, science and technology. Second, he is wondering whether it is really possible to abstract from the entire multiplicity of historical religions something as abstract as the essence of religion—its idea, concept, or essence. Can we extract or filter out from these different religions a smallest common denominator, a single essence that is common to all religions and that unambiguously explains what religion *is*? In fact, both senses of abstraction are problematic because they seem at least partly to presuppose their object of enquiry—a common definition of the concept of religion—in that its essence is seemingly already defined by a single religion, Christianity.

8.4 The Two Sources of Religion and Their Relation to Technology

After this introductory discussion of religion as an abstraction, Derrida continues by asking the following questions: "*Should one save oneself by abstraction or save oneself from abstraction? Where is salvation, safety?*" (§1). He here uses the French word *salut*, which has two meanings: on the one hand, it can mean salvation or rescue; on the other, it can refer to health and security, intactness and immunity. In a suggestive but undeveloped way, Derrida goes on to connect religion with the concept of salvation: "*Can a discourse on religion be dissociated from a discourse on salvation: which is to say, on the holy, the sacred, the safe and sound, the unscathed, the immune (*sacer, sanctus, heilig, holy*, and their alleged equivalents in so many languages)?*"[20] Derrida here questions whether it is really possible to talk about religion without also talking about *salut*, both in the sense of salvation from evil *and* in the sense of health and soundness, the sacred, the unscathed, and the immune.

In fact, the word *salut* has not two but three meanings. While the first two exhibit a certain resemblance, the third has another meaning entirely: *salut* in the sense of a "welcoming" of the other. Derrida thus differentiates between the *salut* that affirms and underpins one's identity, protecting it from evil by providing salvation, health, and security, and the *salut* that involves an unconditional "welcoming" of the other and that challenges one's own identity. In other words, we might say that *salut* has

[20] Every time the word *salut* appears in "Faith and Knowledge", Derrida repeats this long series of words. He does this because the word *salut* has no single equivalent in another language, having so many different meanings and nuances in French, which complement and sometimes even seem to contradict each other. For simplicity's sake, I use whichever word or words fits best in the immediate context.

two fundamental meanings, two origins or *two sources*: in part, the words is used as a substantive in a *constative phrase* such as "the security or salvation of a people" and "the salvation offered by a religion"; in part, the word "*salut!*" operates as a *performative phrase*, a greeting that is *offered* and not only stated and that has its origin in an elementary faith in the wholly other, a faith that finds expression within the framework for every specific historical religion.

In his attempt to get closer to what we call "religion," Derrida thus advances the thesis of religion's two sources, which are referenced in the subtitle of "Faith and Knowledge": on the one hand, religion has its origin in the experience of salvation, health, security, and the protection of the self or of society from external threats, an experience of healing, restoration, or redress; on the other hand, it has its origin in the experience of *faith*, trust, and reliance. When Derrida speaks of "faith" here, he is referring not to the dogmatic faith of the various historical religions but to an originary or elementary faith, an *engagement* in the wholly other—*le tout autre*—without any of the security or promise of salvation or health that he regards as a precondition not only for religion in the traditional sense but also for the social ties that bind society together, for our era's anonymous monetary system with its virtual financial transactions, *and* for technology and science.

After outlining the thesis of religion's two sources, Derrida therefore adduces another thesis, which claims that the same elemental faith underpins both religion and science and technology. He formulates this second thesis about the existence of an essential relation between knowledge and faith, technoscience and faith, as follows:

> *Knowledge and faith, technoscience ... and belief, credit, trustworthiness, the act of faith will always have made common cause, bound to one another by the band of their opposition* (§2).

Technoscience is therefore always essentially bound up with faith and thus religion. For this reason he states in §29 that "one would blind oneself to the phenomenon called 'of religion' or of the 'return of the religious' *today* if one continued to oppose so naïvely Reason *and* Religion, Critique or Science *and* Religion, technoscientific Modernity *and* Religion." Following the thesis that religion has two sources, one being an experience of the sacred or sacrosanct and the other being an elementary faith in alterity, Derrida thus advances the thesis that technology and science also have a basis in this elementary faith, which is to say that religion, technology, and science, despite their obvious differences, have one and the same origin.

It is interesting to note that in 1998, just a few years after the publication of "Faith and Knowledge," Pope John Paul II (a white, European man with a background in the Cracow school of phenomenology founded by Roman Ingarden) issued an encyclical titled "Fides et ratio"—"Faith and Reason." Its primary thesis is that faith and reason are bound up with each other and that reason must rediscover its basis in faith. Although the two texts differ in their understanding of faith and reason, there is clearly a more than superficial resemblance between them.[21]

[21] Naas, *Miracle and Machine*, p. §44.

The statement that religion and technology have a common origin in an elementary faith is not self-evident, however. In a commentary on "Faith and Knowledge" titled "Derrida and Technology: Fidelity at the Limits of Deconstruction and the Prosthesis of Faith" (2002), Bernard Stiegler argues for an alternative position. A superficial reading of the two texts can give the impression that Derrida and Stiegler reach the exact same conclusion with regard to the relation between religion and technology: one is inextricably intertwined with the other. However, closer scrutiny reveals that this unity conceals a deeper and more important disagreement regarding the very basis for Derrida's philosophy of technology. Stiegler refers to a fatal "hesitation" with regard to technology in Derrida's text.[22] According to Stiegler, Derrida's quasi-transcendental logic increasingly resembles a classical transcendental philosophy that brackets the very technology it professes to be welcoming. When Derrida insists that *différance* cannot and must not be reduced to technology, Stiegler reads this demand as a "somewhat surprising declaration of *independence*" with regard to technology.[23] On the contrary, for Stiegler, *différance* is necessarily dependent on technology, which it must, as he puts it, pass "through."[24] It is not enough to argue, as Derrida does, that religion and technology have a common origin—whether that origin is called "God" or *différance*—because any such reference to a quasi-transcendental foundation risks reinstating the very opposing term that it is trying to deconstruct. From Stiegler's perspective, therefore, Derrida's elementary faith is preceded in turn by an even more originary technicity: "*Technics is the condition as much of science and knowledge as of religious faith.*"[25]

As is clear from the discussion in Chap. 3 however, Derrida argues that this position is scarcely tenable: in fact, it is both logically and topologically impossible to reduce the quasi-transcendental level to empirical technology. It is not possible to acquire knowledge of empirical technology when that technology is itself the condition of possibility for knowledge. In contrast to Stiegler, Derrida argues that technology, like religion, presupposes a faith, a trust, a confidence, and so forth—an originary and elementary faith that is the quasi-transcendental condition of all knowledge.

After presenting his thesis that faith is the common origin of religion and technology, Derrida moves on to a discussion of the question that is usually associated with the concept of salvation, namely *evil* and, above all, radical evil. "*Where is evil today, at present?,*" asks Derrida. What are the sites of radical evil, from which humanity in our time seeks salvation, the salvation that is religion's primary source? Derrida finds it in a particular abstraction:

> *We would therefore like to link the question of religion to that of the evil of abstraction. To radical abstraction. Not to the abstract figure of death, of evil or of the sickness of death, but to the forms of evil that are traditionally tied to radical extirpation and therefore to the*

[22] Stiegler, "Derrida and Technology", p. 241.

[23] Ibid., p. 253.

[24] Ibid., p. 260.

[25] Ibid., p. 259, Stiegler's emphasis.

> *deracination of abstraction, passing by way—but only much later—of those sites of abstraction that are the machine, technics, technoscience and above all the transcendence of teletechnology* (§2).

Although Derrida engages in this passage with a central Kantian concept, he tears it from its original context in order to resituate it within contemporary technoscience. Radical evil is not, as Kant maintained, an ineradicable impulse to act on reasons other than moral law; it is the *evil*—the supposed evil—that is always associated with radical abstraction, with extermination, with the delocalized movement that is realized by means of the machines of teletechnology and the virtuality of telecommunications. It was this same radical dislocation, with its roots in delocalization, that I addressed in the previous chapter, in which I summarized Derrida's idea that technological development may eventually confront the world with enormous ethical and political challenges: unbounded mobility of people and cultures, epidemics and capital; uncontrolled waves of migration and flows of homeless and stateless refugees; new forms of terrorism; near-insoluble ethnic, social and religious conflicts; a fast-growing trade in arms and drugs; new forms of colonization; and so forth.

Finally, Derrida presents his third "hypothesis," which I see as the main thesis of "Faith and Knowledge," namely that religion will react to the rootlessness and delocalization of this radical evil by devoting itself to precisely the teletechnological resources that it is simultaneously revolting against. In other words, there is an unavoidable autoimmune relation between religion and technology:

> *In order to think religion today* abstractly, *we will take these powers of abstraction as our point of departure, in order to risk, eventually, the following hypothesis: with respect to all these forces of abstraction and of dissociation (deracination, delocalization, disincarnation, formalization, universalizing schematization, objectification, telecommunication etc.), "religion" is at the same time involved in reacting antagonistically and reaffirmatively outbidding itself* (§2).

It is the relation between salvation and radical evil *as abstraction* that leads Derrida to discover the autoimmune relation between religion and technology. Confronted by the abstracting power of technoscience and telecommunications, its vast capacity to tear things up by their roots, to de-incarnate and delocalize, and so forth, religion reacts antagonistically *at the same time* as it continually embraces this tremendous force, not by passively taking it in but by continually raising the stakes and outbidding itself. In other words, religion reacts against a teletechnoscience and a telecommunication whose resources create rootlessness, homelessness, and delocalization by appropriating those same teletechnoscientific and telecommunicative resources with a view to returning to what those resources threaten: security, unscathedness, one's own language, one's own culture, the nation, the earth, blood, the family, etc. In other words, religion must appropriate the very thing it opposes: our era's most advanced teletechnology and technoscience.

The abstracting power of teletechnology and technoscience are thus always simultaneously bound up with and openly antagonistic towards certain religious values. In my view, this autoimmune relation between technology and religion is the overarching theme of "Faith and Knowledge" from the first page to the last. To be able to speak of "religion" in the singular, religion as such, religion as pure abstraction, we must always relate it to something else, something that lies beyond religion itself—technology.

This *autoimmune* aspect to the relation between religion and technology is also apparent from a later passage in which Derrida declares that there are "at least *two* families, two strata or sources that overlap, mingle, contaminate each another without ever merging … one of the two is precisely the drive to remain unscathed, on the part of that which is allergic to contamination, *save by itself, autoimmunely*" (§28). "The drive to remain unscathed" is *autoimmune* if and when it lays claim to the teletechnological machine, that is, the very machine that has traditionally been regarded as being exterior to religion itself but that our era's "wars of religion" are heavily reliant upon or in which they put their faith. Today more than ever, we know that the modern wars of religion are struggles less over how we should understand the true nature of the sky, the heavens, or the starry firmament than over who should have control of that sky. Today's wars or religion are being fought in *cyberspace*, more or less in real time, *live*, with the help of the very latest teletechnological and media inventions, not as they once were, on earth, with reports of their outcome only being spread around the world after the cessation of hostilities. Digital culture is thus less a resource for reporting on these conflicts than it is the very field of conflict:

> Like others before, the new "wars of religion" are unleashed over the human earth (which is not the world) and struggle even today to control the sky with *finger and eye*: digital systems and virtually immediate panoptical visualization, "air space," telecommunications satellites, information highways, concentration of capitalistic-mediatic-power-in three words (*en trois mots*), *digital culture, jets* and *TV* without which there could be no religious manifestation today, for example no voyage or discourse of the Pope, no organized emanation of Jewish, Christian or Muslim cults, whether 'fundamentalist' or not (§27).

Digital culture, jets, and *TV*—the digitalization of practically all information and communication in the world; the rapid movement of goods and people around the world, and the global spread of television: these three are, in a nutshell, Derrida's image for contemporary efforts to overcome the limitations of time and space. Together, these three are driving the rootlessness, delocalization, and so forth that the first of religion's two sources is seeking to resist by means of the same sophisticated technologies and media.

Derrida provides more examples of this intimate collaboration between our era's religious manifestations and digital culture, rapid communications, and the global media, ranging from the globetrotting of a "Pope versed in televisual rhetoric" to the Rushdie Affair and the growing use of the latest technology by various fundamentalist groups. He describes in an evocative image how "a heavenly glance, monstrous, bestial or divine, something like an eye of CNN" continually watches over us, as if God's all-seeing eye had been replaced by a global satellite network, guaranteeing not only that the entire world can follow all the battles in these new wars of religion but also that the telecommunications network itself is now involved in the frontlines of these wars. This heavenly eye, which is now connected to the world's most sophisticated GPS systems and Google Maps, continually watches over Jerusalem and its three monotheisms, over the multiplicity, the unprecedented speed and scope of the moves of a Pope versed in televisual rhetoric … over airborne pilgrimages to Mecca; over so many miracles transmitted live (most frequently, *healings*, which is to say, returns to the unscathed, *heilig*, holy, indemnifications) followed by commercials, before thousands in an American television studio; over the international and televisual diplomacy of the Dalai Lama, etc. (§27n17).

These examples could be supplemented by a long series of others from earlier periods, from the political influence of televangelists in the United States and elsewhere to the worldwide protests on the Internet and on social media that followed the publication of Lars Vilks's drawings of the prophet Muhammad in the Swedish newspaper *Nerikes Allehanda*. Indeed, this technological theme is so central in "Faith and Knowledge" that Derrida considered giving the essay one of the following titles: "'*Religion and mechane,' 'religion and cyberspace,' 'religion and the numeric,' 'religion and digitality,' 'religion and virtual space-time*'" (§2).

Religion is, and always has been, united with technology in a kind of unholy alliance, but the accelerating pace of technological development in recent years is forcing religion to react by continually raising the stakes in order to catch up and overtake a movement that, in devastating fashion, has perhaps already passed it by. It is thus hardly surprising that Derrida summarizes the principal thesis of "Faith and Knowledge" in the following pithy fashion: "Religion today allies itself with tele-technoscience, to which it reacts with all its might" (§37).

What is really proper (*propre*) to religion—its irrepressible alliance with and incurable allergy to technology—is thus what makes it so fundamentally unspecific. If protecting one's own identity is inevitable, as Derrida claims, then what is specific to religion is both its attempt to protect the first of religion's two sources and its unavoidable transition to something that it specifically is not. It is this paradoxical transition—this autoimmunity—that makes religion use technology in its efforts to protect everything that is traditionally considered more or less sacrosanct: family, home, community, language, nation, nature, and finally what, for Derrida is the most important of all, *life itself*. In the final instance, it is in the name of life that humanity and its communities seek salvation, the life that has thereby been saved, healed, sanctified, made healthy and sound, etc. Religion always seeks to protect and improve life with the help of technological supplements, against which it then reacts automatically like a machine. Or, as he writes in §37: "The reaction to the machine is as automatic (and thus machinal) as life itself." Ultimately, it is not merely religion but life itself that reacts in autoimmune fashion, which is the underlying reason for Derrida's heavy emphasis upon autoimmunity in "Faith and Knowledge." As we recall from the discussion of human technicity in Chap. 5, he means by this that life as a biological process is intimately interwoven with the mechanistic; from a biological perspective, it is a repetitive, automatic, and mechanical process. From the very beginning and in every instant thereafter, life is "a process of self-replacement, the handling-down of life is a *mechanike*, a form of technics." In the same way as the biological process involves a certain death for every conception of life *prior to* the mechanistic, any individual or community who wishes to remain healthy and sound will suffer certain death if they oppose technology. On the contrary, in order to protect the sanctity of life, religion must rely upon the very technoscientific resources that it is resisting with all its might.

8.5 Life, Sacrifice, and Technoscientific Violence

What we call religion in the strict sense of the word—the historical religions—has always had a double, if not aporetical, relationship to life itself: *on the one hand*, an absolute respect and reverence for life, particularly human life, something that is manifested in the commandment "Thou shalt not kill," in sanctions against abortion, artificial insemination, all forms of genetic engineering, and so forth; and *on the other hand*, an equally universal sacrificing of life.

In "Faith and Knowledge," Derrida argues that not only is religion in the service of life, it also sacrifices lives in the name of life because it regards life as having value only as absolute life. Life can only attain its absolute value if it is worth more than itself and only if it can be sacrificed in its own name. It is through this faith in that which is most alive that religion can turn *against* life and ally itself with technoscience—"this enemy of life in the service of life"—in order to defend life itself at all costs. Religion and technology both build on this *sur-viving* logic; they are closely bound up with each other in that they simultaneously serve life and exterminate it, simultaneously believing in the absolute value of life and in its demonization. It is in the defence of the sanctity of life that religion, in accordance with this autoimmune logic, lays claim to the technological resources that it is simultaneously seeking to resist.

We need to analyse this relation between religion and life more closely if we want to understand not just religion's efforts to protect life at all costs, indeed, even at the price of life itself, but also the spectral life that necessarily emerges from this sacrifice. Derrida develops his line of thought about this spectral life in a long passage in §37 in which he points to several possibilities or forms for what he calls "the return of the religious" in our time. One such form for the return of the religious is the "self-destructive affirmation of religion" in its globalized form, that is, the autoimmune reaction which exists in all of the three monotheistic religions, namely their "pacifist" appeal for a "universal fraternization" that seeks to go beyond and thereby exterminate all established religions in the name of a kind of worldwide "ecumenical reconciliation." What these pacifist, humanitarian, and ecumenical attempts to go beyond religion signify is that a divine order is simply replaced or sacrified in the name of the infinite value of the human.

According to Derrida, the return of religion, if such a thing exists, need not necessarily be measured in numbers of church visits but can also be manifested in various cosmopolitan, ecumenical, pacifist, or humanitarian movements, which, despite superficially seeming secular or non-religious, in fact follow the same autoimmune logic as traditional religion. Accordingly, Derrida here claims that the autoimmune rejection of technoscience by means of technoscience can follow two avenues that compete with each other and are apparently antithetical. Both of them, however, can as easily oppose or support a "democratic" tradition: either the fervent return to national citizenship (patriotism of the home in all its forms, affection for the nation-state, awakening of nationalism or of ethnocentrism, most often allied with Churches or religious authorities), or, on the contrary, a protest that is universal, cosmopolitan or ecumenical: "Ecologists, humanists, believers of all countries, unite in an International of anti-tele-technologism!" (§46).

As sanctifying as it is sanctimonious, this protest against teletechnology by means of technology provides an effective illustration of Derrida's thesis about the autoimmune character of religion as well as of the central place of life—the sanctity of life—in religious discourse in our own era.

This life must be held in absolute respect, treated with dignity, and, as we have seen, protected at all costs, including at the price of life itself. In other words, life can assume two opposed forms: an absolute respect for life and its sacrifice. The former respects and protects life, as, for example, in the commandment "Thou shalt not kill" (if not those alive in general, then certainly not those around you), while the latter, the "universal sacrificial vocation," not only protects life but seeks to exceed it by sacrificing it. For example, this sacrifice of life is today practiced in the form of industrialized farming and slaughter of animals as well as animal experiments; they are literally *sacrificed* in the name of human life. This practice of animal sacrifice follows a *mechanical* principle which "is apparently very simple: life has an absolute value only if it is worth *more than* life." In other words, it cannot be reduced to biological life but must continually exceed itself, continually go beyond the purely biological. Respect for life thus applies only to human life, the only life that in any way attests to "the infinite transcendence of that which is worth more than it (divinity, the sacrosanctness of the law)" (§40).

This theme of sacrifice, which recurs frequently in Derrida's writing, from *Glas* (1974) and *Donner la mort* (1992) to "Un ver à soie" (1998), is always related to the mechanical, the technical, the technological. Respect for life and a willingness to sacrifice are two postulates, he argues, that follow on from each other with the regularity of a technology or a machine, in which death lies at the very heart of processes whose purpose is to protect and even enhance life. The fact that life is worth more than life is a *mechanical* principle for the very reason that respect for life and a willingness to sacrifice "reproduces, with the regularity of a technique, the instance of the non-living or, if you prefer, of the dead in the living" (§40). Note how Derrida's line of argument here connects to his definition of technology as the relation between life and death, between the living and the non-living. Death and life, "death in the living," are not two opposed and independent entities; with the regularity of a *technology*, they always already presuppose each other.

Derrida builds his argument by first connecting the dignity of life to the concepts of transcendence, fetishism, and spectrality, all of which are associated with belief in the other, that is to say, with the very core of religion, but which play a new and unprecedented role in religions that are today defined by teletechnoscience:

This dignity of life can only subsist beyond the present living being. Whence, transcendence, fetishism and spectrality; whence, the religiosity of religion. This excess above and

> beyond the living, whose life only has absolute value by being worth more than life, more than itself-this, in short, is what opens the space of death that is linked to the automaton (exemplarily "phallic"), to technics, the machine, the prosthesis: in a word, to the dimensions of auto-immune and self-sacrificial supplementarity, to this death-drive that is silently at work in every community ..., constituting it as such in its iterability, its heritage, its spectral tradition (§40).

The life that only has absolute value by being more than life, more than life itself, can thus only be protected or saved by that which is associated with death: technology, the machine, the prosthesis, and the specific machine that Derrida here takes up—the phallus. Derrida devotes the larger part of §39 to discussing the example of the phallus or what he calls "the phallic effect." In order to understand what he means by these concepts, it is useful to return to the principal thesis of "Faith and Knowledge" since the phallus or the phallic effect is perhaps the best example of the way in which the teletechnological supplement, in autoimmune fashion, seeks to protect life, the life that is worth more than life itself. The phallus, he argues, is the technological supplement of pure presence itself, "its ghost, its spectre, its double or its fetish" (§39). The phallic effect is the "*colossal automaticity* of the erection (the maximum of life to be kept unscathed, indemnified, immune and safe, sacrosanct), but also and precisely by virtue of its reflex character, that which is most mechanical, most separable from the life it represents." The phallic effect thus unites, simultaneously and paradoxically, life in its purest form, untainted by any machine, *and* the machine as a mechanical supplement without autonomy that nonetheless seems more living than life itself. As a mechanical supplement, the phallus is "distinct from the penis and once detached from the body," nothing less than "the marionette that is erected, exhibited, fetishized and paraded in processions." The phallus is a symbol of power, an essential attribute for sovereign power, for the state's power over life and death, or quite simply for the potency of the political leader. The use of sophisticated space technology that guarantees a dominant political, economic, and militarily strategic position of power in the world is perhaps the most striking and visible expression of this phallic effect, which unites "the tele-technoscientific machine, this enemy of life in the service of life," to "the very source and resource of the religious: to faith in the most living as dead and automatically *sur-viving*, resuscitated in its spectral *phantasma*, the holy, safe and sound, unscathed, immune, sacred-in a word, everything that translates *heilig*" (§39). It is in defence of the sanctity of life that religion, in accordance with autoimmune logic, lays claim to the technological resources that it is also trying to resist. And fundamentally, it is autoimmune logic that is the reason why religion cannot be separated from violence and war.

Perhaps the monstrous crimes that are being and have been committed in the name of various fundamentalist movements are not the only things that can be defined as wars of religion. Perhaps the same also holds true of certain wars that are being and have been fought in the name of secularism, enlightenment, and reason. "Wars or military 'interventions' led by the Judaeo-Christian West in the name of the best causes (of international law, democracy, the sovereignty of peoples, of nations or of states, even of humanitarian imperatives)—are they not also," asks Derrida, "from a certain viewpoint, wars of religion?" (§28). In order to determine whether a war is a war of religion, we need first to be able to delimit the religious. For Derrida, however, this is impossible: the religious is not confined to religious belief in the traditional sense; it exceeds the absolute opposition between faith and knowledge, between religion and technology, and works by using the technology that it also opposes.

According to Derrida, the *violence* of wars of religion fall into two categories. The first is the *contemporary* violence that is strongly associated with ultra-efficient military teletechnology—a digital cyberculture. The second is a new, *archaic* violence that "counters the first and everything it represents" (§42). It revenges itself by using the same tele-technological and mass-media resource to attack the naked body and pre-mechanical life. It retaliates by committing beheadings with bare hands, castrating, mutilating, raping, and exposing defenceless bodies, all in order to return to the dream of the sanctity of life: "What is involved is always avowed vengeance, often declared as *sexual* revenge: rapes, mutilated genitals or severed hands, corpses exhibited, heads paraded, as not to long ago in France, impaled on the end of stakes (phallic processions of 'natural religions')" (§42). And, Derrida adds with an eye to the complete absence of women from the Capri conference, this is perhaps the place to ask why women are always the most defenceless victims of the appalling violence that is necessarily ethnic or religious in nature (§39).

As Derrida sees it, archaic violence is a revenge upon a global teletechnoscience, a reaction to economic and military hegemony and to a European, democratic model that is simultaneously secular and religious (§42). It is this autoimmune reaction to teletechnology's expropriation and delocalization, which, in his view, can explain the seemingly paradoxical alliance between the contemporary wars of religion and "irrationalist obscurantism with hypercritical acumen and incisive analysis of the hegemonies and the models of the adversary." Derrida predicted that archaic religious violence would escalate and become increasingly ghastly. Wars, which in some sense are also wars of religion, will unite the most advanced technoscience with a reactive barbarism that seeks to directly attack its own body, what Derrida calls "the *sexual* thing," which, if nothing is done, will continue to be raped, mutilated, or desexualized. Unfortunately, events have proven him right. In order to do justice to this senseless violence, we will need to integrate *psychoanalysis*, not only "the most powerful discourses today on right [i.e. law], morality, politics, but also on science, philosophy, theology, etc." (§43).

8.6 Religion and Media

We have seen how Derrida identified religion's two sources: an experience of salvation on the one hand, and on the other an elementary faith in the entirely other, a faith that simultaneously constitutes the basis for all science and technology. But we have discussed, above all, his thesis that religion must always be understood in relation to technology—particularly today—and to the teletechnology that defines our world to such a profound degree and that is so fundamentally changing our understanding of it.

In this section, I focus not on the relation between religion and technology in general but more specifically on the relation between religion and media. In "Faith and Knowledge" and his other later writings, Derrida argues that religion today is indistinguishable from the media that spread religious messages across the world via books, radio, the Internet, and, above all, the medium that interested him perhaps most of all: *television*.[26]

Let us return to one of Derrida's many writings on this subject, the filmed conversation between him and Bernard Stiegler, which is recounted in concentrated form in the book *Échographies de la télévision*, which we examined in detail in Chap. 4. This improvised conversation was not Derrida's first TV recording. Indeed, he was a regular guest in TV studios around the world and was an avid consumer of television. In answer to Stiegler's transparently conversation-making question about which TV programmes he liked to watch, Derrida replied: "Aside from the news, I watch all kinds of things, the best and the worst. Sometimes I watch bad soap operas, French or American, or programs that give me a greater cultural awareness … political debates, spectacular political encounters in general … or else old movies. I could spend 24 h a day watching good political archives. … And so I watch a little of everything. … What few people I know watch regularly, I suspect, and I watch very regularly Sunday morning, from 8:45 to 9:30, are the Muslim and Jewish religious programs, which I find very interesting—and if we had time to talk about them, I'd tell you why."[27]

Well, why is Derrida so interested in watching religious programmes on French television? We find the answer in "Faith and Knowledge" amongst a series of improvised commentaries from December 1997.[28] It turns out that Derrida's highly regular viewing habits on Sunday mornings reveal something more than an ecumenical or multicultural interest in the world's great religions: they reveal a long-term personal interest in the production of the programmes themselves, their medialization, and their dissemination of their religious message. What primarily interested him was not the religious content of these programmes but the relation between religion and global media, particularly television and its unique role in the reproduction of the religious message. But the really interesting thing about Derrida's ostensibly unphilosophical reply to an unphilosophical question turns out to be his opinion that only Christianity ought to be called a religion and that the globalization of media should be understood as a primarily Christian phenomenon.

In an enlightening passage in *Miracle and Machine*, Michael Naas claims that the key to the differences that exist, as Derrida sees it, between the three monotheistic religions with regard to the relation between religion and media is their view of *secrecy*.[29] Secrecy seems to constitute an invisible, inseparable, unquestionable core

---

[26] We can only speculate about how interested Derrida would have been in the social media of our own era, but my guess is that the more user-defined and two-way nature of Facebook, Twitter, YouTube, etc. would have interested him every bit as much as television did.

[27] Derrida & Stiegler, *Échographies de la télévision*, p. 155.

[28] These comments were later published as an essay titled *Surtout pas de journalistes!* (2005).

[29] Naas, *Miracle and Machine*, pp. 127–34.

of religion, an originary faith in the wholly other, one of religion's two sources. In *Donner la mort* and other works, Derrida has addressed the question of secrecy, particularly as it is represented in the narrative of the bond between God and Abraham, a narrative that is broadly common to Islam, Judaism, and Christianity.[30] According to Derrida, the testing of Abraham consists not only of God's command that he sacrifice his own son but also the necessity of not revealing the secret that God has entrusted to him, the necessity of not reporting what He has said or commanded. As Derrida reads it, God's requirement that Abraham not reveal what has passed between them in secrecy must "remain absolutely secret: just *between us*. It must remain *unconditionally* private, an internal affair and inaccessible." In other words: "Above All, No Journalists!"[31] What God really wanted to tell Abraham—in fact, what he actually said to him—was this: reveal nothing, do not spread, missionize, or preach about what has passed between us; do not make our secret into good news that will be reported in the news, and, above all, do not spread it via television, mobile phones, or—Derrida would now probably have added—the whole gamut of social media.

This understanding of secrecy allows Derrida in *Surtout pas de journalistes!* to begin to sketch some important differences between the three Abrahamic religions, something that in turn raises doubts about the possibility of identifying something as abstract as the essence of religion. Derrida claims that Abraham's secret "is the major reference for Judaism and Islam" in a way that it is not for Christianity, which has essentially replaced the central dogmas about Abraham's secrecy and the interrupted sacrifice of a son with the good news of God's own son who sacrificed himself in order to become the mediator between humanity and God.[32] Derrida further argues that this difference explains both the ban on images in Judaism and, in particular, Islam and their relative reluctance to spread religious messages via the media, that is to say, via a *medialization* of word and image. Unlike Judaism and Islam, Christianity seems to regard its *essential* task as being to spread the "good news" across the world, to communicate it globally via books, radio, TV, satellites, the Internet, and so forth. It is for this very reason that Derrida dares to boldly conclude that medialization as such is "fundamentally Christian and not Jewish, Islamic, Buddhist, etc." Although medialization is a phenomenon that occurs "in all religions," there is "a trait that is absolutely singular in the power and structure of Christian medialization." It is this unique quality that Derrida, using a somewhat awkward neologism, calls *mondialatinization*, which can perhaps be translated as "globalatinization" and which expresses the idea that globalization is inseparably bound up with both Christianity and the language through which Christianity was spread across the world, namely Latin (and whose role in our own era has passed to English).

---

[30] For a deeper analysis of the connection between God and Abraham, see, for example, *Donner la mort*, p. 101–3.

[31] Derrida, *Surtout pas de journalistes!* p. 8.

[32] Ibid., pp. 10–1.

For Derrida, then, globalization in the form of teletechnology and medialization *in itself* seems to be a Christian phenomenon, while it is a contingent phenomenon for Judaism and Islam. It is *still* a Christian phenomenon when Judaism and Islam apply its strategies and technologies. This means that the category of religion, with its Latin roots and with the publicity and medialization that have emerged from it and that have been promoted by it, are a fundamentally Christian concept.

From this perspective, religion seems to be synonymous with Christianity. Christianity seems to be the only religion, the only doctrine or religious practice that can be called religion. Even if it is tempting at this point to object to Derrida's exclusion of Judaism, Islam, Buddhism, and so forth from the category of religion, we can perhaps agree with his secondary point, namely that we must be careful to call other "religious" practices or doctrines *religion*, because we then—by including them, in a spirit of generosity and ecumenism, among the world's great religions—are perhaps already "globalatinizing" or Christianizing them.[33]

According to Derrida, for a religion to gain the status of a world religion, it must be engaged in a "globalatinized" or universal process that is essentially Christian.[34] In other words, other religions exist only insofar as they can appear in an essentially Christian space. Derrida's entire argument is thus founded on the existence of a number of different religions on the basis of which he tries to extract a common essence, before finally finding that only one religion defines this essence. The entire time his thinking moves between singular and plural, individual and universal, religion's multiplicity and its claim to universality:

> All these religions are doubtless religions with a universal vocation, but only Christianity has a concept of universality that has been elaborated into the form in which it today dominates both philosophy and international law. There is in St. Paul a concept of cosmopolitanism, a concept of world citizen, of human brotherhood as children of God, etc., which is closer to the concept of universalism as today it dominates the philosophy of international law than are other figures of universalism. … Thus one has to distinguish very precisely the values of universality that are at the heart of the three religions called monotheistic. The universality that dominates global political-juridical discourse is fundamentally Greco-Christian. Or at least I believe it is. It is a Christianity speaking a bit of Greek.[35]

All religions unquestionably strive for universality, Derrida underscores in this passage, but it is Pauline Christianity's special variant of this concept that has attained such authority and legitimacy across the whole world and that has come so profoundly to define the international conversation, international law, and the international institutions.

Derrida regards this Pauline universalism as deeply ambivalent: it is both promising and problematic. On the one hand, it seems "very mild" in inviting all of humanity, as God's children, to be part of the new community; on the other, it appears "terrible" and violent if and when it seeks to impose the totalitarian idea of one

---

[33] Ibid., p. 19.

[34] Ibid., p. 49.

[35] Ibid., pp. 49–50.

humanity united in a single faith.[36] According to Derrida, this deconstructive fluctuation between two impossible alternatives is a hallmark not only of Christian universalism but also, and to the same degree, of the mediality with which it is essentially bound up.

Christianity, alone among all the world's religions, seems to be essentially connected with the concept of universality and mediality. As Derrida also states: "All the Christian churches are more mediatic than their Jewish, Moslem, Buddhist, etc. equivalents."[37] For this reason, he can refer, a few lines later on, to "the televisual hegemony of the Christian churches," particularly in the United States, where the message of Christianity is spread via television in a dramatic, spectacular, and highly effective way, primarily by the Baptist and other evangelical movements but also by the Catholic Church. The question is: what is really going on in the United States and why is it about this aspect of American television that so interests Derrida? The object of his fascination seems to be television's capacity to fascinate, its capacity to ostensibly give access to the thing in itself—the Biblical miracle—*live*, without technological mediation.

In addition to the Biblical miracle, there is another kind of miracle, the most common of all miracles: "believe me!." When I say "believe me!" to someone, I am appealing to an elementary faith that lacks any kind of evidence, a faith that can never know what is really going on in the mind of the other. Faith like this, which exceeds all knowledge, is the same as believing in a miracle. By exploiting this elementary faith, which Derrida regards as the basis for all media, by shamelessly exploiting this unreserved "believe me!," which is the precondition for all communication and all social ties, Christianity seeks to combine the television image's seemingly "real presence" with the production of Biblical miracles. The media thus exploits both our elementary faith in the social bond *and* our increasing faith in teletechnology and media as a way to produce miracles. It is upon this "reduction of faith" that media establish their power: "Television and media exploit this reduction of faith … Nothing will be left but the barest foundation of the social bond."[38] When we talk these days about the end of the great ideologies, the death of philosophy, and the end of all ethical-political systems, all that remains is the bare foundation of the social bond—our originary or elementary faith.[39]

Today, Derrida argues, it is primarily religion, in an alliance with teletechnology and media, that continues to exploit this originary or elementary faith. In other words, it is on the basis of this fundamental act of faith and this experience of the miracle that teletechnology can create the impression that the miracle is actually happening *live*. In "Faith and Knowledge," he therefore refers to all these "miracles [being] transmitted live (most frequently, *healings*, which is to say, returns to the unscathed, *heilig, holy*, indemnifications) followed by commercials, before

[36] Ruin, "Circumcising the Word: Derrida as a Reader of Paul", p. 92.

[37] Derrida, *Surtout pas de journalistes!* p. 14.

[38] Ibid., pp. 28–9.

[39] Ibid.

thousands in an American television studio" (§27n.17). In *Surtout pas de journalistes!*, he puts the matter even more graphically:

> Even if masses are not generally celebrated, "miracles" are shown. Some of you will have seen in the United States such miracles on the screen: bodies falling backward on the order of the officiating thaumaturge or rising once cured, etc. One is truly in the Bible, seeing paralytics walk, the blind see—all of that, naturally, "live," before the eye of the camera, all organized by powerful industrial corporations, with formidable quantities of capital at their disposal.[40]

It is this capitalization of elementary faith—which is the precondition for teletechnology—that makes possible the production of miracles *live* on the TV screen. This would not have been possible, Derrida notes, without a comprehensive teletechnological knowledge and, above all, gigantic capital investment, which together create a problematic connection in American religious programmes between preaching and marketing, prayer and product promotion. Although these programmes can be seen as an extreme form of Christian evangelizing via media, the fact remains, he argues, that Christianity as such is essentially connected to the concepts of universality and medialization.

For Derrida, this essential connection between Christianity and media is ultimately based on incarnation, that is to say, the idea that God assumed a human form in the person of Jesus Christ. The death of God, he argues, occurred with the incarnation and Jesus's death on the cross. No longer in heavenly isolation, untouched by worldly forces, God thereby ceases to be a living reality and is instead brought within the time and space of those worldly forces. For Derrida, the death of God becomes the real central idea of Christianity, an idea that continually returns in different ways in the Christian tradition, all the way up to Hegel's "absolute knowledge" as the goal of world history in *Phenomenology of Spirit* and Friedrich Nietzsche's famous slogan "Gott ist tot" in *The Gay Science*.[41] Yet Christian media are not merely effective as means for spreading the idea of God's death around the world; they are the very milieu in which this idea takes shape and is produced:

> This Christian hegemony in the world of televisualization (*dans la mondia-télévisualisation*) is the hegemony of a religion founded on the ordeal of *kenosis* and of the death of God. Terrifying mystery. The Christian message remains tied to the incarnation, the death of God, under the form of Christ on the Cross giving his body, but also, *as a result*, through all the deaths of God that have followed in the history of European philosophy until Hegel and

---

[40] Ibid., pp. 16–7.

[41] Derrida's thesis of "the death of God" as Christianity's real central idea bears a striking resemblance to the "God-is-dead-theology" that has been formulated by a group of radical theologians whose best-known exponents are Thomas Altizer, Paul van Buren, William Hamilton, and Gabriel Vahanian. Thomas Altizer is particularly interesting in this context because his theological writings cite Hegel and Nietzsche (whom he characterizes, somewhat curiously, as radical Christians). Like Derrida, he argues that "the death of God" is God's own self-effacement, his negation of himself in Jesus. In Jesus, the transcendental and holy God has died once and for all and "risen again" in humanity's immanent and profane development process. (Altizer & Hamilton, *Radical Theology and the Death of God*, 1966).

> Company. What propagates itself as media, as Christian telemedia, is also a kind of death of God.[42]

In other words, there is an extremely close connection between the Christian message and everything going on in our own era. This connection would be impossible without the doctrine of *kenosis* and God's death, without a certain "atheism," without the idea of "Christianity *as the end of Christianity*."[43] We are today perhaps experiencing the end of a Christianity that uses media better and more effectively than all other religions, and the beginning of a *religion of the media* that is both Christian and post-Christian. Derrida seems to have taken this thesis from Hegel's *Phenomenology of Spirit*:

> At the end of the *Phenomenology of Spirit*, it is the figure of Absolute Knowledge, the structure of philosophy as the truth of revealed religion, which is to say of the Christian religion. The "Speculative Good Friday." In abusing these shortcuts … I would venture to say that the *religion of the media* is a *phenomenology of spirit*. Today there is a *religion of the media*, and it is, for the moment, precisely *the religion of the media*: there is the religion *of the* media, which is to say, the religion that the media represent, incarnate, reveal.[44]

Religion, which, according to Derrida, is essentially Christian, thus seems to have died along with Jesus on the cross, only to "rise again" in our own era as a religion of the media, a world-historical event signifying salvation; it is perhaps thus the gospel of our time, our era's "good news." This "good news" has now reached every corner of the Earth, thanks to the universality of media and their hitherto unknown power. While a certain "analogue" technology has always existed, what is happening today—the accelerating development and continual refinement of teletechnology and media, globalization, the dramatic transformation of time and space, and the digitalization of essentially everything that can be digitalized—all this is an effective universalization of religion without precedent in the history of humanity. Never before have religion and technology been so closely intertwined—locally, regionally, and globally. And never before has technology had such power over the human mind and such *redeeming* power. In our own era, the path to salvation increasingly reveals itself in the shape of technology.

8.7 On Atheism and Deconstruction

Finally, let me return to the current debate between John Caputo and Martin Hägglund over the so-called religious "turn" in Derrida's later writings. Without going into the various rounds of the debate, which is sprawling and full of misunderstandings on both sides, I want to explain how they view deconstruction and its relation to technology.

---

[42] Ibid., p. 34.

[43] Ibid., p. 38.

[44] Ibid.

In sharp contrast to the prevailing view of a religious "turn" in Derrida's writing, Martin Hägglund argues in his book *Radical Atheism: Derrida and the Time of Life* (2008) that Derrida's work, from first to last, is characterized by a radical atheism. Atheism has traditionally confined itself to denying the existence of God and human immortality while not questioning the *desire* for God and immortality. By contrast, radical atheism claims that desire for immortality in fact conceals a more originary longing for *survival* that contradicts it from within.[45] Instead of putting our hopes in a transcendent *beyond*, everything we do presupposes an engagement with and care for our spiritual state in the here and now.

Hägglund argues, quite simply, that the concept of survival is incompatible with immortality because life is essentially transient and by its nature subordinate to the laws of time. Every moment in life is about survival because it is dependent on the temporal structure of the *trace*. As we recall, this structure requires that every now must be inscribed in space as a trace just in order to continue living. The trace makes it possible to preserve the past because the former is characterized by a capacity to continue living despite the passing of time. The trace is thus the fundamental condition of life; it can only live on if it is given up to a future that may erase it.

For Hägglund (and Derrida), time and space are not—as in Kant—transcendental forms of human intuition that are given in the same way, independently of empirical conditions. On the contrary, their constitution is dependent on the empirical or the material-technological: "if time must be spatially inscribed, then the experience of time is essentially dependent on which *material supports* and *technologies* are available to inscribe time" (emphasis added). The inscription of time in space therefore explains why "there is neither a beginning nor an end to historicity and technicity."[46] For Hägglund, life is essentially bound up with the technological.

Hägglund also emphasizes that "autoimmunity is a condition for *life in general.*"[47] When religions try to protect life and make it absolutely immune, they continually run the risk of being subjected to an autoimmune reaction. As Derrida argues in "Faith and Knowledge," it is in the defence of the sanctity of life that religion, in accordance with autoimmune logic, lays claim to the very technological resources it is trying to resist.

John Caputo countered Hägglund's book with an article in *Journal for Cultural and Religious Theory* titled "The Return of Anti-Religion: from Radical Atheism to Radical Theology" (2011). In it, he welcomes the book as an "impressive exercise in deconstruction." But despite being frank about its "strategic advantages," he is less impressed by its substance. According to Caputo, Hägglund's book argues for a certain *kind* of deconstruction, an abbreviated version of Derrida, a deconstructive torso adapted for a younger generation of unreasonable materialists: "All scrubbed

---

[45] Hägglund, *Radical Atheism*, p. 1.

[46] Ibid., p. 27.

[47] Ibid., p. 14.

up and sanitized, nothing written in the margins, deconstruction as logic, not *écriture*."[48]

How does Caputo himself understand deconstruction? On the basis of Derrida's definition of deconstruction as "the experience of the impossible," he develops his own "radical theology" according to which deconstruction is a religion *without* religion. This does not mean that he understands religion in terms of a supreme being called God; on the contrary, he follows Derrida in denying that the impossible *is* God. God is neither the supreme reality or the pure act (*actus purus*) but "the name of an event in which we find ourselves at risk, exposed to what drives us to the limits, to what we desire with a desire beyond desire, to what we love *per impossibile*, with prayers and tears."[49] For Caputo, God is not a being in the hereafter, who does or does not do impossible things, but "a limit concept, a dream."[50]

At a seminar with Derrida held at Stockholm's Södertörn University in 2000, this limit concept came up for discussion. Interestingly, Derrida described his relationship to the wholly other—*le tout autre*—as precisely a "dream." The idea of peace, he said, "is somewhere in my heart, in my hope, and it is what structures my experience." Hägglund, who also participated in the seminar, replied with surprise: "I don't understand how dreaming can be at the heart of deconstruction ... I can't see a role for a dream." To which Derrida responded: "No, you are right, and that is why I don't take this dream too seriously. But I dream the dream!."[51]

Derrida "dreams" the impossible: peace, justice, and the democracy to come. It is in this specific sense that he (and Caputo) can refer to deconstruction as a religion without religion—a purely formal structure of waiting, expectation, and hope in contrast to a theological dogma.

Caputo describes his philosophy of religion as a materialist theology or a religious materialism, an approach that he claims is today widely accepted and that Hägglund refuses to acknowledge. Yet he welcomes Hägglund's argument for a *certain kind* of materialism in Derrida and the way in which he clarifies that *différance* is not an immaterial spirit but "requires a *material substrate*" (emphasis added). He also shares Hägglund's view that there is no other place where deconstruction can "take place" in time and space. Caputo imagines that this was the background to Derrida's idea that Husserl, by explaining "the origin of geometry" in terms of "writing," was able to claim that the constitution of ideal objects "requires a material-technological substrate … the dualism of ideal and material, soul and body, *physis* and *techne*."[52] For the same reason, it undermines the dualism between religion and the material: religion is rather a particular way of understanding the

---

[48] Caputo, "The Return of Anti-Religion", p. 33.

[49] Ibid., p. 39. Caputo here refers to Derrida's admission in *Circonfession* that, by virtue of being a "little black and Arab Jew … [he] has been praying all his life, kissing his prayer shawl every night."

[50] Ibid., p. 37.

[51] Derrida, "Från lag till rättvisa" ["From Law to Justice"], pp. 118–9.

[52] Ibid., s. 65. Caputo argues that it is just as inadequate to say that Derrida is a materialist or a realist as it is to say that he is an idealist. The least confusing option would be to say that he is not an "anti-materialist or an anti-realist, even as he is not an anti-idealist".

material in our world. Caputo's religious materialism accords neatly with Derrida's thesis about a religion of the media, particularly when he argues, towards the end of his essay, that "the classical figure of the 'angel' (*angelos*, messenger) is gradually instantiated by today's advanced information technologies."[53]

For me, Caputo is most convincing when he embraces Hägglund's interpretation of deconstruction as a materialist "logic" while also claiming that it is only one of several possible interpretations—deconstruction, as Derrida states in *Donner la mort*, can only be called logic "up to a certain point."[54] Although Hägglund's reading of Derrida is also convincing in many respects, it seems one-sided in denying what he calls religion in Derrida in order to make way for a radical atheism that cleanses deconstruction of all trace of transcendence and that expels religion of every kind from Derrida's texts.

From a certain perspective, we might say that Caputo's argument is the mirror image of Hägglund's. Whereas Hägglund wants to use Derrida in order to free us from religion once and for all, Caputo seems to want to associate him with a largely harmless kind of religion that has been liberated from all the historical baggage of religious violence. One wants to make us immune to all forms of religion; the other wants to make us immune to all impulses to religious violence. As I see it, both readings are one-sided. Indeed, with his desire for survival, Hägglund seems to have unintentionally drawn inspiration from religion's first source, the desire for immunity for one's own identity. In similar fashion, Caputo seems to emphasize religion's second source, faith in the wholly other in the absence of any security or promise of survival. This same one-sidedness recurs when Hägglund asserts that "radical evil" is more originary than the "good" that Caputo prefers to highlight—in reality, Derrida continually asserts that both are *equally originary*. Trying to ward off radical evil or longing for good, meeting the wholly other—*le tout autre*—with fear and trembling or with love, is ultimately undecidable.

Even if Hägglund and Caputo, from a certain perspective, seem like mirror images of each other, they are in agreement on at least one point: there is a close connection between deconstruction and technology. We cannot distinguish between a purely "religious" and a purely "technological" deconstruction because all experience is necessarily mediated by technology. Moreover, religion and technology are always already intertwined in an autoimmune relation. The drive to defend, protect, and save all that religion considers sacred—family and home, community and nature, language and nation, even life itself—this inexorable drive is autoimmune when it lays claim to the teletechnological machine, the machine that is traditionally thought to lie outside of religion itself but which it so profoundly relies on and puts its trust in. Ultimately, it is in defence of the sanctity of life that religion, in accordance with autoimmune logic, lays claim to precisely the technological resources that it is also trying to resist.

---

[53] Ibid., p. 120.

[54] Ibid., p. 47.

8.8 Summary

In this concluding chapter, I have tried to sketch the main features of Derrida's philosophy of technology as a whole. I began by briefly presenting some of the central concepts and ideas discussed in the preceding chapters and then devoted the main portion of the chapter to his views on religion.

Derrida's life was shaped in different ways by his Jewish, Muslim, and Christian background. He often refers to his deep interest in questions of religion, even as he considers himself "atheist" in the sense that he does not believe in the existence of an other-worldly supreme being, a metaphysical-philosophical superego. After taking an interest in negative theology in the 1980s, which he gradually came to see as a dead end, he began in the 1990s to develop a deconstructive theology that can be summarized in the paradoxical phrase "religion without religion." Its stated goal was to liberate theology from its metaphysical-philosophical superego, revealing instead what he saw as an authenticity in the gospel—the messianic openness to the wholly other, *le tout autre*—that comes from an unforeseeable future. It is this so-called religious "turn" that Derrida's successors would interpret in such diametrically different ways. Most regarded this religious turn as the logical consequence of the very core of deconstruction, but others saw it as an illegitimate attempt to inscribe a transcendent discourse into this core. Yet, as I have sought to show, both these readings are one-sided: we cannot distinguish between a purely "religious" and a purely "technological" deconstruction because religion and technology are always already connected in an autoimmune relation.

The question of the relation between religion and technology is not new. It has been on philosophy's agenda since at the least the beginning of modernity, even if it has traditionally been described as an opposition between religion and science, faith and reason, faith and knowledge. The central questions in "Faith and Knowledge" are about precisely this relation between faith and knowledge, religion and science, religion and technology. Having reflected in several previous chapters upon the conference's *staging*, a strategic move that he regards as necessary for an understanding of something as abstract as religion, Derrida goes on to ask the question: what *is* religion? How, he asks the conference delegates, can we six European men, gathered here on Capri, isolated from the world, all with a clearly Christian background, talk about religion? How can we, *in this special situation*, without representatives from all the other religions, speak of religion, religion in the singular, without generalizing or perhaps even profaning? How can we speak of religion at all when the language we use, perhaps from the very beginning, is the language of religion, in other words a Latin language, which shapes so many of our religious concepts, including the concept of religion itself? To ask what religion *is* by using the word "religion" itself is problematic, to say the least, because the essence of religion is perhaps already defined by one particular religion—Christianity.

Derrida has a genuine respect for the question of religion and even refers to his fear and trembling at it. In his efforts to nonetheless get to grips with the question, he advances what can be summarized in three theses. The first states that religion has two sources: on the one hand, salvation, security, and protection of the self or society from external threats; on the other, faith, trust, and confidence. When Derrida refers to "faith," he has in mind not the dogmatic faith of the various historical religions but an originary or elementary faith that, as he sees it, is a precondition not only for religion in the traditional sense but also for the social ties that bind a society together *and* for technology and science.

Following the thesis on religion's two sources, Derrida therefore presents a second thesis, which claims that this same elementary faith forms the basis for religion as well as for science and technology. As a result of this faith, technology is essentially connected with religion.

The inevitable tension between religion's two sources—protection of one's own identity and faith in the wholly other—leads Derrida to advance a third thesis, which, I argue, is also his main thesis: the relation between religion and technology is not only essential but *autoimmune*. This autoimmune relation between religion and technology perhaps shows itself primarily in religion's attempts to use technology to protect life, the sanctity of life. What we refer to as religion in the strict sense of the word—the historical religions—has always had a double, if not aporetical, relationship to life itself: *on the one hand*, an absolute respect and reverence for life, particularly human life; *on the other hand*, an equally universal sacrifice of life.

It is easy to agree with Derrida's statement that religion not only is in the service of life but at the same time sacrifices lives in the name of life because, for religion, life only has value as absolute "life." Life can only attain its absolute value if it is worth more than itself and only if it is sacrificed in its own name. It is by means of this "faith in the most living" that religion can turn *against* life and ally itself with technoscience in order to try to defend life itself at all costs. It is out of a reverence for the sanctity of life that religion, in accordance with this autoimmune logic, lays claim to the very technological resources that it is also trying to resist.

In the final section, I take up the specific relationship between religion and media. As Derrida argues, religion today is indistinguishable from the media that spread the religious message across the world through books, the Internet, radio, and television. His main claim here is that the medialization of the world is fundamentally a Christian phenomenon. Medialization may occur in all religions, but there is one aspect that is completely unique to Christianity: the desire to spread the "good news" of the gospel across the whole world by means of media.

For Derrida, this essential connection between Christianity and media is ultimately founded on incarnation. The death of God occurred with the incarnation and Jesus's death on the cross. For Derrida, the death of God becomes the real central idea of Christianity, an idea that continually returns in different ways in the Christian tradition, all the way up to Hegel's "absolute knowledge" as the goal of earthly processes in *Phenomenology of Spirit*. We are today perhaps experiencing the end of a Christianity that uses media better and more effectively than all other religions, and the beginning of a *religion of the media* that is both Christian and post-Christian. Perhaps religion died along with Jesus on the cross and "rose again" in our own era as a religion of the media, a world-historical event signifying salvation and thus perhaps the gospel of our time, our era's "good news." This "good news" has now reached every corner of the Earth, thanks to the universality of media and their hitherto unknown power. Christian media have achieved an effective universalization of religion that is without precedent in the history of humanity. Never before has technology had such power over the human mind. In our own era, the path to salvation increasingly reveals itself in the shape of technology.

References

Altizer, T., & Hamilton, W. (1966). *Radical theology and the death of god*. Bobbs-Merrill.

Caputo, J. D. (2011). The return of anti-religion: From radical atheism to radical theology. *Journal for Cultural and Religious Theory, 11*(2), 32–12.

Derrida, J. (2009). *La contre-allée. La Quinzaine Littéraire*. Louis Vuitton.

Derrida, J. (2005a). *Surtout pas de journalistes!* Éditions de L'Herne.

Derrida, J. (2005b). Från lag till rättvisa [From Law to Justice] (trans. N. Smith). In *Lagens kraft [Force of Law]*. Symposion.

Derrida, J. (2001). Faith and knowledge. In G. Anidjar (Ed.), *Acts of religion*. (trans. S. Weber). Routledge.

Derrida, J. (1999). *Donner la mort*. Galilée.

Derrida, J. (1985). Deconstruction in America: An Interview with Jacques Derrida. By J. Creech, P. Kamuf, J. Todd. *Crititical Exchange, 17*, 1–33.

Derrida, J., & Stiegler, B. (2002). *Echographies of television: Filmed interviews*. (trans. J. Bajorek). Polity Press.

Derrida, J., & Stiegler, B. (1996). *Échographies de la télévision*. Galilée.

Habermas, J. (1987). *The philosophical discourse of modernity*. (trans. F. Lawrence). The IT Press.

Hägglund, M. (2011). The radical evil of deconstruction: A reply to John Caputo. *Journal for Cultural and Religious Theory, 11*(2), 126–150.

Hägglund, M. (2008). *Radical atheism: Derrida and the time of life*. Stanford University Press.

Hart, K. (2004). Religion. In J. Reynolds & J. Roffe (Eds.), *Understanding Derrida* (pp. 54–62). Continuum.

Naas, M. (2012). *Miracle and machine: Jacques Derrida and the two sources of religion, science and the media*. Fordham University Press.

Ruin, H. (2013). Circumcising the word: Derrida as a reader of Paul. In P. Frick (Ed.), *Paul in the grip of the philosophers: The apostle and contemporary continental philosophy*. Fortress Press.

Stiegler, B. (2001). Derrida and technology: Fidelity at the limits of deconstruction and the prosthesis of faith. In T. Cohen (Ed.), *Derrida and the humanities*. Cambridge University Press.

Chapter 9
Conclusion

In this study, I have shown that Derrida's thinking in all its different phases has returned to the question of technology and the technological as fundamental philosophical questions. His various treatments of them collectively form a coherent reflection that I have sought to reassemble and highlight. Taken together, these reflections constitute an appropriate framework for a philosophy of technology, a *phenomenology of technology*, that is neither transcendental nor empirical. In fact, we can trace the roots of this phenomenology of technology back to the early works of Martin Heidegger. In *Being and Time*, as is well known, Heidegger describes human existence's way of being-in-the-world. He views tools and instruments as the central feature of the relation between humans and their world. Tools and instruments are what shape the human encounter with the world. The world reveals itself to humans most immediately in that which is "ready-to-hand" rather than that which is "present-at-hand". What lies "at hand" are precisely the instruments or tools that we use in our everyday activities.

This early initiative on Heidegger's part could well serve as the basis for a phenomenology of technology that takes different *technologies* seriously. Unlike these earlier initiatives, however, his later works of philosophy of technology suffer from three "difficulties". First, Heidegger sometimes tends to downgrade *techne* to a secondary role in relation to a pre-technological originary state—*physis*. For Heidegger, *physis* seems to represent something resembling a truth that is independent of *techne*. His questioning of a modern lifestyle characterized by rampant consumption, superficial values, and exploitation of the natural world undoubtedly seems anti-technological. Second, he sometimes seems to want to view humanity's originary essence and identity as purely spiritual and as untainted by the technological. This "pure" form of spirituality is problematic, if we accept Derrida's thesis that human life in all its dimensions is always already contaminated by the technological-material. Third—and this is perhaps the most significant difficulty—Heidegger does not manage to develop an adequately hermeneutical perspective on technology, since he reduces technology to its conditions of possibility—its essence—and

B. Sjöstrand, *Derrida and Technology: Life, Politics, and Religion*,
https://doi.org/10.1007/978-3-030-83407-4_9

thereby avoids interpreting its relation to concrete material technologies. For Heidegger, there is a sharp line of demarcation between technology's essence and technology itself, which is intended to "purify" thought of essence of any technology.

Heidegger's view of modern technology as an all-encompassing, independent, and alienating force, which follows its own logic and that holds society and culture in its grip, is largely shared by most phenomenologists and transcendentally-oriented philosophers of technology. For example, Jacques Ellul has written about the *autonomy* of technology that has caused humans to lose control of Technology (with a capital T), something that seems to represent a new way of being-in-the-world. A variant on this theme, albeit of older provenance, is Lewis Mumford's romantically coloured vision of a decentralized, small-scale, and life-affirming "democratic" technology, in contrast to the technology that he sees as dominating today—a centralized, top-down-ruled, and oppressively "authoritarian" technology. Another example is Ortega y Gasset's existentially-oriented philosophy of technology, which holds that modern technology *as such* involves a new mode of operation for consciousness, a mode that ultimately leads to tremendous problems for modernity, namely that humanity's originally creative capacities will wither and disappear.

Although Derrida has not commented directly on the writings of these three philosophers of technology, it is clear that their "transcendentalism", which situates technology's conditions of possibility, rather than technology itself, at the heart of philosophy of technology, is incompatible with his own quasi-transcendental philosophy of technology.

Derrida's philosophy of technology is equally incompatible with the empirical school's research into technology. The "empirical turn" in philosophy of technology represented by Bruno Latour and his theory of "actor networks" has unquestionably increased our understanding of what concrete technology actually *do* in different contexts. Yet the phenomenological, hermeneutical, and existential questions that are so central for Heidegger and Derrida are noticeable by their absence in Latour's work.

It was as an attempt to bridge the polarity between transcendental and empirical philosophy of technology that Don Idhe and Peter-Paul Verbeek launched postphenomonology, which regards subject and object as not only intertwined in an intentional relation but as also *constituting* each other. In light of this approach's interest in studying specific technological apparatuses in people's everyday lives, its strong ties to American pragmatism, and its robust critique of technology's critics, we might nonetheless ask: to what extent does postphenomenology differ *in practice* from a purely empirical view of technology?

Derrida's philosophy of technology has a great deal in common with postphenomenology, not least its resistance to the notion of a radical essence. Notwithstanding these similarities, they differ on several key points. First, Derrida claims that the technological apparatus is a *supplement* to the human subject. It is an external "memory", and as such it is a necessary supplement that living human memory needs in order to function in the first place: technology is supplementary but not constitutive. Human consciousness is not constituted by empirical technology—as Verbeek, for example, seems to think—but by a quasi-transcendental *différance* that cannot be reduced to empirical technology. Second, Derrida's phenomenology of technology is not only critical of technology's critics, it is equally critical of the seemingly neutral, not to mention uncritical, attitude to technology adopted by postphenomenologists and others. For Derrida, technology is neither good nor bad but *both* good and bad. Third, postphenomenology seems to be primarily interested in using empirical research to study the technological apparatuses themselves, "things in themselves" and "what things do" (to quote the title of Verbeek's perhaps most famous book), unlike Derrida, who engages not only with these apparatuses and their practical applications but also, and above all, with the accelerating developments within various fields of technology and their respective ethical, political, and religious consequences.

Derrida often refers to a world-encompassing teletechnological apparatus, a "writing machine" that mediates knowledge of the world across distances not only in space but also in time. The accelerating development of this teletechnological apparatus has undeniably had a dramatic effect on ethics, politics, and religion as well as on how we humans conceptualize time—a time that is not only being reduced at a steadily increasing rate but that has, in fact, gone quite some way towards losing its original character of a homogeneous temporal flow.

Derrida is deeply engaged with this process, which he regards as a fundamental disturbance of the human experience of time. His own conception of time proceeds from Husserl's transcendental philosophy, which is based on the inner temporalization of consciousness. He replaces this transcendental logic with a line of thinking founded on *différance*. On this view, time is manifested in the form of *inscription* in the space that is opened up as *writing* and in the form of an originary temporal *synthesis* whose movement is maintained by the inscription of writing itself and that constitutes the condition of possibility for all meaningful experience of time. This arche-writing—or *différance*—thus replaces Husserl's transcendental logic as the framework for the production of meaning.

At the same time, Derrida connects *différance* as a framework for creating meaning with the concept of *avenir*, that is to say, with the other, the wholly other, which suddenly—as if by chance—comes from an uncertain and unforeseeable future. Yet *l'avenir* should be seen not as something that actually occurs at a future point in time but, rather, as something that intervenes with messianic force and interrupts the homogeneous flow of time here and now. What thereby emerges is a complex and qualitatively more heterogeneous time, one that cannot be thought of as linear but must instead be seen as twisted, a time that cannot be understood in terms of "before" and "after". *This time is out of joint.*

This dislocated time becomes dislocated precisely by virtue of being produced artificially. Time is an *artifact* that is produced by the teletechnological apparatus. To take an example, television never mediates an originary event *live*, as pure presence; it necessary produces *pseudo-events* by editing, slanting, and formatting "information". According to Derrida, a pseudo-event produced in this way is inscribed into the very heart of the event structure at the same time as it necessarily differs from the originary event. It is the ethical task of deconstructive thinking, not to confuse appearance with reality, but to analyse the pseudo-event by remaining attentive to the originary event.

Yet Derrida's understanding of time is more radical than this. For him, not only these pseudo-events but *all* events are technologically mediated because all events must be inscribed as *traces in space*, for which reason the human conception of time is essentially dependent on the material aids and technologies that are available for inscribing time. Traces, these representations of technology, constitute time though movement in *différance*. Derrida's time is therefore incompatible with the notion that time is constituted upon an ultimate foundation, whether that foundation takes the form of transcendental subjectivity, existential temporality, or empirical technicity.

Différance is the condition of possibility for the human experience of time—and not only for this experience but for experience as a whole. Derrida even goes so far as to claim that *différance* as the condition of possibility for human experience is nothing less than life itself, life in general. *Différance* is life. Through movement in *différance*, life—human life as well as life in general—is therefore always already *contaminated* by technology. Life is not technology, but nor is it the opposite of technology; from a certain perspective, we must acknowledge that a certain kind of "technology", such as genetic code, is "inscribed" in life in a way that is analogous to (although clearly not identical with) writing in the conventional sense. *Différance* is the "programme" of life and, as such, encompasses the entire chain of life, from the genetic inscription of the first living cell, through the emergence of human beings, and all the way to alphabetical writing and our own era's inscription in digital archives. If Derrida is right in claiming that *différance* is life in the general sense of the word, then the appearance of human beings is "merely" one stage in the history of life. In other words, there is a continuity in the "programme of life", ranging from genetic inscription to alphabetical writing and beyond.

Derrida's understanding of humanity as in a state of continual formation differs from Bernard Stiegler's idea that human beings are merely "inventions". The key to the invention of the human is precisely the role he assigns to *technology* in human evolution. The latter is not merely a process of biological evolution, it is also an evolution of technological aids without which human beings would not exist. For Stiegler, human history is a history of technology in which the technological invents the human and the human invents the technological. As he sees it, the invention of technology by humans, the externalization of humans in technology, is a unique event in the history of life. At that precise moment something fundamentally new began, namely a shift from "cerebral cortex to flint", from genetic to non-genetic. For Derrida, by contrast, this shift is continual.

What Derrida regards as the continual shift from genetic to non-genetic presupposes that nature and technology, *physis* and *techne*, are not each other's polar opposites but in fact always already intertwined. This necessary intertwining means that "technology" is inscribed not only in human life but in life in general, including genetics. Genetic inscription is admittedly not technology in the usual sense of the word, but it can nonetheless be thought of as a kind of "technology" insofar as it does connect the living with the non-living in general: the DNA molecule in fact connects a living organism to its non-living predecessors to exactly the same degree as the written or technological legacies of those predecessors. What is more, genetic inscription is not a one-time event but a repetitive, mechanical process. As Derrida notes, it is a *mechanike*, a form of technology.

Although life, from a certain perspective, must be described in technological terms, it does not mean that life *as such* is a form of technology. Derrida does not subscribe to a biological reductionism that views humans as purely technological-biological beings. Conversely, however, it is not possible to reject this genetic version of biologism without also embracing some kind of metaphysical subjectivism. A strategy of counterposing biologism to one-sided subjectivism inevitably results in our being forced to choose the "least dangerous" of the two. The alternative strategy proposed by Derrida involves developing Husserl's transcendental-phenomenological concept of subjectivity into what I have called "scriptual" subjectivity.

This elaboration of Husserl's phenomenological discourse means that the subject can no longer be thought of as the centre of being but must instead be described as an "effect" within a system of writing that is open to the technological world around it. The psyche or psychical apparatus operates like a "writing machine" that continually inscribes and erases traces from this larger technological world. Consequently, we can no longer treat the psychical apparatus as the *polar opposite* of a non-psychical apparatus: in reality, these two apparatuses are increasingly coming to resemble each other. There is a clear analogy between living human memory and the present era's complex systems of inscription and storage of almost unlimited amounts of data in external memories. It is for this reason that Derrida can define the concept of technology as, not an apparatus, but a relation between two apparatuses, between the psychical and the non-psychical, between life and death.

To summarize: the unique feature of Derrida's *alternative* philosophy of technology—his deconstructive phenomenology of technology (a version of philosophy of technology)—would seem to be its extension of the concept of technology to encompass the entire phenomenological field: there is no possibility of shielding our thinking about essence from empirical technology. In our very desire to shield ourselves from this technology, we are more exposed to the risk of being "contaminated" than if we try to conceptualize technology in terms of *contamination*. Technology is not a polar opposition between essence and supplement but a relation, a contamination, in which being and supplement presuppose each other. We must presume this contamination, Derrida argues. Without it, we will be unable to take responsibility for technological development and globalization. Without it, we will be unable to remain open to the wholly other, to the messianic, to the future with all its threats and possibilities.

It is ultimately tempting to ask the question of what technology really *is*—this technology that, according to Derrida, "is always already there", all the way down to the first living cell. What *is* Derrida's technology, which is inscribed in everything from the amoeba and the annelid to the zeroes and ones of our own era's information and communication systems? Paradoxically, technology is *nothing*. Neither presence nor absence, neither being nor supplement, it is the very play of essence and supplement, of presence and absence, the opening of the game that no metaphysical or ontological concept can comprehend. For this reason, the younger Derrida proposes that a new name must perhaps be found in order to remove it from its traditional problematic—and, we might add, resituate it within the infinitely slender space between life and death.

The manufacturer's authorised representative in the EU is Springer Nature Customer Service Centre GmbH, Europaplatz 3, 69115 Heidelberg, Germany. If you have any concerns regarding our products, please contact ProductSafety@springernature.com

Printed and bound by CPI Group (UK) Ltd, Croydon, CR0 4YY
28/09/2026
02224088-0007